Distillation in
Nontechnical Language

DISTILLATION

IN NONTECHNICAL LANGUAGE

Norman P. Lieberman

PB
PENNWELL
BOOKS

Disclaimer

The recommendations, advice, descriptions, and methods in this book are presented solely for educational purposes. The author and publisher assume no liability whatsoever for any loss or damage that results from the use of any of the material in this book. Use of the material in this book is solely at the risk of the user.

PennWell Books, LLC
10050 E 52nd Street
Tulsa, Oklahoma 74146 USA

866.777.1814
sales@pennwellbooks.com
www.pennwellbooks.com

Publisher: Matthew Dresher

Cover image: Getty Images (ID: Walid Shemeis)

Library of Congress Cataloging-in-Publication Data available upon request.

Identifiers:

LCCN 2026008226 (print)

ISBN 9781955578363 (hardcover)

ISBN 9781955578370 (eBook)

Printed in the United States of America

1 2 3 4 5 30 29 28 27 26

Dedication

To the light of my life, Liz, my wife, who shares both the Chemical Engineering profession and my work in designing and troubleshooting refinery process equipment. I would be lost without her.

To my beloved daughter, Irene, who guides me through the later stages of my life, and is responsible for converting my scribbled notes into the text of this book.

To the 22,000 attendees of my 1,020 seminars, who have tolerated my abrasive style of technical presentation since 1984, but still email me with process questions.

To Henry Kister, my coconspirator in the application of process technology to hydrocarbon distillation problems, ever since 1980, when we started talking about my first book.

Contents

Chapter 32

Preface

This book was written for professionals responsible for operating, maintaining, and designing process equipment. Refineries, hydrocarbon processing facilities, natural gas liquefaction plants, and petrochemical facilities should only be operated, maintained, and designed by people who are dedicated to operating, troubleshooting, designing, and optimizing the process equipment for which they are responsible.

The qualifications that have best helped me to understand how distillation towers work are to have:

- Served in the military
- Grew up on a farm
- Worked as a plumber
- Paid attention in high school science class
- Tried to make firecrackers as a child, from a chemistry kit

Troubleshooting distillation equipment is a complex and fascinating occupation, one that I've been practicing since I graduated from Cooper Union in New York, in 1964, as a chemical engineer.

Designing and operating distillation towers has always been my main activity. But the most rewarding aspect of my 60-year career has been troubleshooting, which has meant crawling through the tower tray deck manways during a turnaround, discovering tray assembly errors, instructing maintenance as to what needs fixing, and, finally, seeing a successful result when the tower is re-streamed. That's what makes it all worthwhile.

Purpose of a Distillation Tower

The purpose of a distillation tower is to separate two or more components, such as water and alcohol. The water, which boils at a higher temperature, will come out of the bottom of the tower. The alcohol, which boils at a lower temperature, comes out of the top of the tower.

To reduce the alcohol content of the water, heat is added to the bottom of the tower, to generate vapor, to strip most of the alcohol from the water that leaves the bottom of the distillation tower.

To reduce the water content of the alcohol, heat is removed from the top of the tower, to generate liquid (called reflux). This washes the water out from the rising alcohol vapors.

The more heat that is added to the bottom of the tower, and the more heat that is removed from the top of the tower, the better the separation will be between the alcohol and water.

The reflux washes out water from the rising vapors. The stripping vapors strip out alcohol from the falling liquid.

The liquid and vapor mix on a perforated plate called a "Tray." The liquid runs across this perforated tray, and overflows a "Weir" at the edge of the tray into a "Downcomer," and then onto the tray below.

The better the mixing, the more efficient the tray is. The greater the number of trays, the better the separation is between the alcohol and water. The more reflux running down the trays, the better the separation is between the alcohol and water. The more vapor that rises through the trays, the better the separation will be between the alcohol and water.

Distillation was invented 900 years ago in Persia (now Iran) for the purpose of making a beverage with high alcohol content (i.e., brandy). Modern technology has not altered the basic concepts of distillation, and never will! Whether we are separating:

- Propane from butane,

or

- Kerosene from diesel oil,

or

- Gasoline from jet fuel,

the technology, as I have described it in this text, will apply until the end of mankind's tenure on Earth.

1

THE ORIGIN OF DISTILLATION TECHNOLOGY

Conversion of Wine into Brandy

This chapter traces progress made by mankind over time to increase the alcohol content of fermented grains and grapes, from 10% to 50%. The leadership in this progress was centered in the Middle East, an area of technical developments long before Europe and America assumed the world's leadership. The technology I will describe is based on a visit my wife and I made to a primitive fermentation and distillation plant in a rural area of Peru that fermented corn to produce a 70% to 80% proof alcoholic beverage for local consumption.

The Basic Idea of Distillation

The concept of distillation is one of the fundamental building blocks of human civilization. It is the ability to separate a mixture of different liquids into individual products and the most common examples are:

- The separation of alcohol and water to produce whiskey and wine.
- The separation of natural gas liquids to produce LPG (propane) and butane.
- The separation of crude oil to produce gasoline, jet fuel, and diesel oil.
- The separation of tar, to produce medicinal mineral oils, lubrication oils, roofing tiles, and paving asphalt.
- The separation of aromatics, to produce chemical-grade benzene, toluene, and xylenes.
- The fractionation of paraffinic waxes.

How Distillation Works

The feed to a distillation tower is a mixture of different liquid components that boil, or vaporize, at different temperatures. These liquids are completely soluble in each other. Wine is a mixture of 13% alcohol and 87% water. If you leave a glass of wine on a table, the alcohol and water will not separate. Instead, you will have to partially separate them using distillation technology, which is what this book is about: How to separate two or more liquid components, based on their individual boiling points or vapor pressures.

The alcohol content in wine has a lower boiling point than the wine. The water content in wine has a higher boiling point than the wine. To separate the alcohol and water in the wine from each other, we use this boiling point difference as our tool.

Vaporization of Wine

If wine is partly vaporized (see Figure 1.1), the resulting vapor will be a mixture of alcohol and water. But, because the alcohol is twice as volatile as the water, the alcohol content of the vapor leaving our boiling pot will be higher than the water content of the vapor leaving the pot. This increase in the concentration of the alcohol in the vapor leaving our pot, relative to the concentration of the water in the vapor leaving the pot, is a function of the "Relative Volatility" of alcohol vs. water. It is this concept of relative volatility that forms the basis for distillation technology.

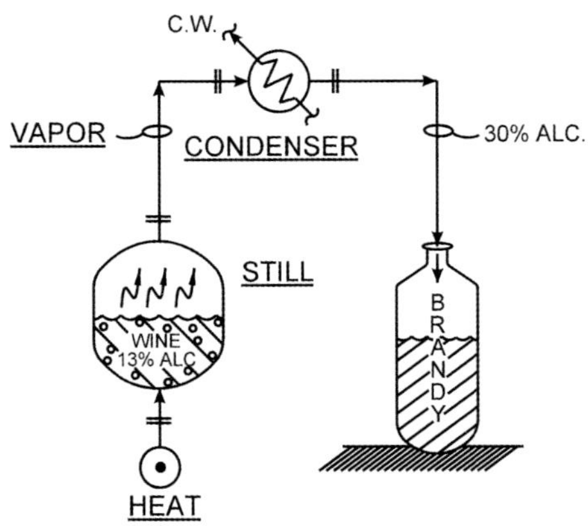

Figure 1.1. Still

It is this concept, that the more volatile components in a mixture of liquids will vaporize more readily from the liquid phase and into the vapor phase, that forms the fundamental basis for all distillation technology.

Partial Condensation

Once the alcohol-rich vapor is generated, by partially vaporizing the wine, there is another step that can be taken to increase the percent of the alcohol in the vapor recovered from the wine: partial condensation of the vapor (as shown in Figure 1.2). Water will condense more readily from a vapor than does alcohol, because water is much less volatile than alcohol. Thus, this uncondensed vapor will be richer in alcohol than the vapor just generated from boiling the wine.

If I then condense this vapor, leaving the partial condenser shown in Figure 1.2, I will have greatly increased the alcohol concentration of my 13% alcohol in the wine, to perhaps 50% alcohol in my brandy bottle, as shown in Figure 1.2. Of course, I will have a lot less brandy to drink, compared to my initial amount of wine, which was fed into my distillation tower.

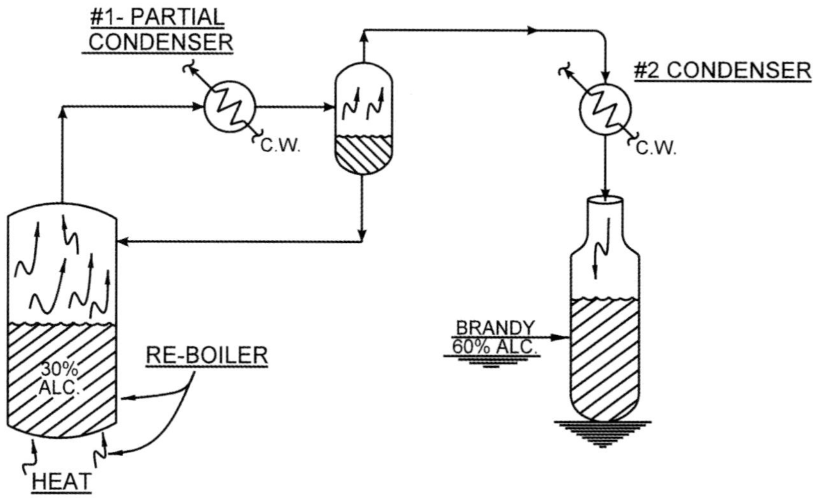

Figure 1.2. Partial condenser

Distillation Technology

The development of distillation technology was based on the objective to increase the alcohol concentration of fermented grains, grapes, and berries, and later to recover the propane, butane, gasoline, and kerosene from crude oil, using the same technology.

The distillation technology as used in petroleum refineries was developed gradually using distillation trays, refluxing, and reboiling over a period of about 900 years. This advance in distillation technology resulted in the "Patent Still" of the 1880s. Since that time, this technology has not changed much (see Figure 1.2).

Function of Reflux in Distillation

My objective is to separate water from alcohol. I will use a distillation tower for this purpose. The wine is the feed to my distillation tower, and as a liquid, it trickles down the trays and is partly vaporized at the bottom of the distillation tower by a fire (or steam, or some other source of heat), which is called the "Reboiler."

The trays consist of perforated plates. The up-flowing vapor that is generated by the reboiler's heat strips out the alcohol (along with some water) from the distillation tower's feed. These plates are called "Distillation Trays." The greater the amount of heat from the reboiler, the more completely the alcohol is removed from the waste water stream called the distillation tower bottoms product.

The vapor mixes repeatedly with the descending liquid to strip out the residual alcohol from the liquid flowing down the distillation tower. When the vapor leaves the top tray, it flows into a water- or air-cooled heat exchanger, called a "Condenser." The condensed liquid, which is mostly alcohol, is divided into two streams:

1. Reflux,
2. Brandy, which is the overhead product that we wish to recover.

The reflux now trickles back down the trays to come into contact and mix with the up-flowing vapors, to wash the water out from the up-flowing vapors rising through the trays.

Function of the Reboiler

The heat input to the tower serves two functions:

1. To strip out the alcohol from the descending water-rich liquid,
2. To generate vapors that will turn, after a few moments, into the reflux.

In order to understand how a distillation tower works, one must visualize that the "REFLUX COMES FROM THE REBOILER."

If something happens to diminish this heat input to the tower, the first effect observed by the distillation tower operator is that his reflux drum will go empty. This is the most critical concept of distillation technology to be understood.

Function of Trays

Each tray is intended to be a vapor-liquid mixing device. The more efficient this mixing, the better the tray efficiency. If the mixing efficiency is perfect, and all of the evolved vapor separates completely from the evolved liquid, leaving the tray, we say that the tray has a "100% Fractionation Efficiency."

In reality, trays never have a 100% vapor-liquid mixing efficiency. At best, it only reaches 80% to 90%. And, sometimes, if the trays are not installed level, their vapor-liquid mixing efficiency may only be 30% to 50%, which is a very common problem!

The number of trays in a typical distillation column ranges from 20 to 40. However, I have designed distillation columns with only ten trays or as many as 70 trays, depending on the degree of difficulty of separation of the lighter and heavier components in the feed. The more difficult it is to separate the two components:

- The more trays we need
- The more reflux we use
- The greater the reboiler heat input or duty

When mankind first began to dominate our planet, he discovered that, when yeast ferments the sugar content of grape juice into alcohol, the yeast runs out of sugar to ferment when the alcohol content of the wine is only about 15%. Even if we add sugar to the fermenting mixture, once the alcohol content of the wine reaches 30%, the alcohol becomes toxic to the yeast in the fermenting mixture. So, if we want to bring brandy to 120 proof (60% alcohol), we have to employ distillation technology, as described below.

1. Assume alcohol has a vapor pressure in our still of 40 psia (pounds per square inch), at the operating temperature in our still shown in Figure 1.1.
2. Assume that water has a vapor pressure in our still of 20 psia.
3. Then, the mole % of vapor that is leaving our still will be 67 mole % alcohol.

$$(40) \div (40 + 20) = 67\%$$

The mole % of both components is the same in the still, but the vapor pressure of the alcohol is twice that of water.

But there is yet another problem that led mankind to develop distillation technology 900 years ago. Assume that we have a still that has 30 mole % alcohol, plus 70 mole % water. As the alcohol-rich vapor leaves the still, the concentration of water remaining in the still will go up.

As the mole % of water in the still increases, the percentage of water in the vapor leaving the still will also increase. And the mole % of alcohol in vapor leaving the still will decrease. Then, the concentration of alcohol in our brandy bottle will go DOWN. This is bad! In other words, the more wine we try to convert into brandy, the lower the alcohol content in the brandy!

Balance Between Product Purity and Product Recovery

If we are satisfied to produce a small amount of brandy, we can make its alcohol content quite high. If we want to maximize brandy production, its alcohol content, or its proof, will be much lower. The objective of distillation technology is to overcome these restrictions.

There are three developments that allowed mankind to overcome limitations posed by his environment:

1. The wheel
2. Production of copper
3. The partial condenser

A sketch of the partial condenser is shown in Figure 1.2. Rather than condense all of the vapor from the boiler, only half of the vapor is condensed in the first condenser (Condenser #1). The water vapors, which condense at a lower temperature than the alcohol vapors, will preferentially condense first. The liquid from this partial condensation step will be mainly water, but with some alcohol. This liquid was called "Flux" by the ancient Persians who developed this technology in what is now Iran.

The other portion of the vapors that were not condensed in Condenser #1, shown in Figure 1.2, will be rich in alcohol, and quite lean in water. These alcohol-rich vapors were then put through a second-stage condenser (see Figure 1.2), which drained into our waiting bottle of brandy (120 proof, 60% alcohol).

But there was still a problem! What to do with the flux, that is, the liquid from the first condenser? At first, it was just discarded into the sands of Persia. But then the very first Chemical Engineer suggested that the flux be drained back into the boiler.

Did this help produce more brandy?

No, not at first.

But then this first Chemical Engineer said, "Let's add more heat to the boiler, to 'revaporize' the flux!" And thus were born distillation and our Chemical Engineering profession.

The Distillation Tower

Currently, we call the flux recycled back to the boiler "reflux," and we call the heat added to the reboiler "reboiler duty." The reflux is generated in the overhead condenser. The reboiler duty is generated in the fired heater, or the steam-heated bottoms reboiler.

Improving Fractionation

The more heat we remove from our first-stage condenser to generate more reflux (Figure 1.1), and the more heat we add to our reboiler, the harder we work our distillation tower. And the greater the recovery will be of our product.

All I have done is to describe a simple distillation tower, with a single theoretical separation stage. Are such distillation facilities still employed?

Absolutely, yes!

I saw one just last year in the front window of a liquor store in Vilnius, Lithuania, where I was troubleshooting a 120,000 BSD Crude Unit distillation tower with my wife, Liz, who is also a Chemical Engineer.

The Problem of Entrainment

As our first Chemical Engineer increased reflux rates to purify his brandy, he noted that, at some point, the separation efficiency of his still was getting worse rather than better. Sampling the vapor leaving the still and flowing into the condenser, he observed that droplets of liquid were blowing out of the still into the condenser. He concluded that these droplets contained a lot of water. This was contaminating the first condenser with water, and therefore increasing the amount of water in the vapor phase that was leaving this first condenser, and thus reducing the alcohol content of the vapor (and the entrained liquid) leaving the first condenser (see Figure 1.3).

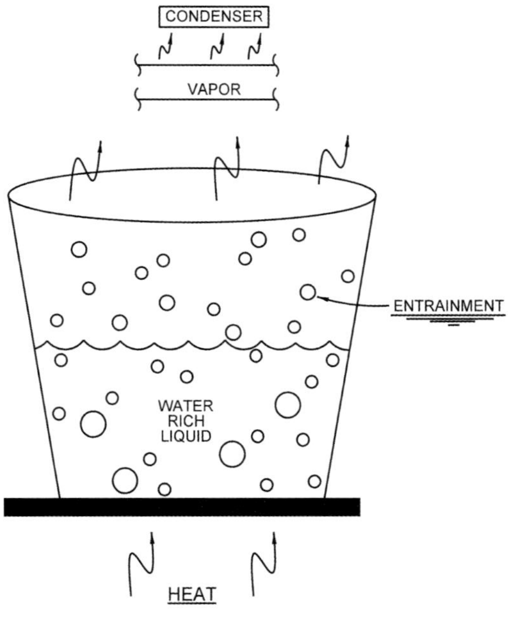

Figure 1.3. Entrainment

Our first Chemical Engineer then concluded that more reflux and more reboiler duty was good, up to a certain point. And that point was when the vapor velocity leaving the still is high enough to promote the entrainment (which is now called "Jet Flood") of the water-rich liquid phase in the still.

Entrainment is the single largest problem that degrades distillation, or fractionation efficiency. As the vapor flow and vapor velocities increase, the liquid spray height being lifted up the distillation tower degrades the fractionation, or distillation, separation efficiency.

As reflux rates increase, so does the vapor flow velocity in the distillation tower, and so does the rate of entrainment (see Figure 1.3).

Summary

The invention of distillation technology, including the introduction of using reflux to greatly increase the concentration of alcohol in wine, allowed mankind to overcome a limitation in creating alcoholic beverages that had persisted for thousands of years. This basic concept of reflux is at the heart of every process plant, and of every refinery, along with the concept of re-contacting the up-flowing vapors with the descending liquid reflux on distillation trays. The production of gasoline, LPG, and lubrication oils would not be possible without the concepts of trays and reflux.

2

VAPOR PRESSURE – ALCOHOL VS. WATER

Alcohol has a vapor pressure that is higher than that of water. If this was not true, then we could not use distillation to increase the alcohol content of wine and convert the wine into brandy.

It would be easy to separate alcohol and water if they were not mutually soluble. It's easy to separate oil and water, because they don't appreciably dissolve in each other. But, alcohol and water are completely soluble and can't be separated by allowing a mixture to settle.

The way we can separate alcohol and water is to take advantage of their difference in "Vapor Pressure." What is the meaning of the vapor pressure of a liquid? Let's say I have a bottle of pure water. No air is dissolved in the water. It's just water. If I heat the water to 212°F (100°C), the pressure of the water in my bottle will rise to 14.7 psia (i.e., atmospheric pressure). If I cool the water in my bottle to 100°F, the pressure of the water in my bottle would drop to 1.0 psia. (Note: psia refers to absolute pressure (pounds per square inch)).

That's the meaning of the vapor pressure of a pure liquid.

Pure alcohol has a vapor pressure that is higher than that of water, at the same temperature. Alcohol's molecules evaporate more easily than water. Let's assume that, at a particular temperature, the vapor pressure of alcohol is three times that of water.

Meaning of Relative Volatility

The ratio of the vapor pressure of alcohol, divided by the vapor pressure of water (which is 3), is:

(Vapor Pressure Alcohol) ÷ (Vapor Pressure Water) = "Relative Volatility"

The higher the relative volatility, the easier it is to separate the two components by distillation. The difference in the vapor pressures of all liquids, such as alcohol and water, ALWAYS increases at lower pressures.

This means that it's easier to separate different components, such as alcohol and water, at a lower distillation tower pressure. This pressure is a variable that can be adjusted on most distillation towers.

Vapor Pressure Charts

You can see that what I have noted is true by looking at a "Vapor Pressure Chart" for hydrocarbon components. The vapor pressure lines spread apart as the temperature is reduced. This means that lower tower pressures make it easier to separate two components because their relative volatility is increasing.

Mole Percent of Components in Liquid

There are two factors that determine the percentage of alcohol in the vapor that is leaving a liquid-filled container:

1. Relative volatility of alcohol vs. water (vapor pressure of alcohol, divided by the vapor pressure of water, at the same temperature).
2. Mole % water and mole % alcohol in the liquid. To determine the mole % of each component in the liquid in our container, we divide:

(Wt. %) ÷ (Molecular Weight)

This determines the relative mole % of each component in the liquid phase. That is, their relative concentration.

To determine the relative amounts (in mole %) of the water vs. alcohol in the vapor leaving our container, we multiply:

- The vapor pressure of each component in our container by the mole percent of each component in our container.

(Vapor Pressure) • (Mole Percent)

As we evaporate more of the liquid in our container, the concentration of water that is left in the container goes UP! And the concentration of the alcohol in our container will go DOWN! As I recover more of the wine into my bottle of brandy, the alcohol content of my brandy is going to diminish, because I am trying to maximize my production of brandy. And, as I try to recover more brandy from my wine, the brandy's alcohol percentage must go down.

The solution to this quandary was distillation. It required civilized humans many centuries to develop the technology that I will describe in great detail in the following chapters.

The Partial Condenser

Referring to the condenser shown in Chapter 1, this is a "Complete or Total Condenser." It is complete in the sense that all of the vapor is condensed into liquid. Therefore, the composition of vapor from the still is the same as the composition of the liquid leaving the condenser and flowing into the brandy bottle.

As early mankind discovered, partial vaporization of wine did generate an increase in the alcohol content in the evolved vapor stream. But, not to the extent that they desired. To overcome this less than desired alcohol content of their brandy, the ancient Persians invented the partial condenser, as shown in Figure 2.1.

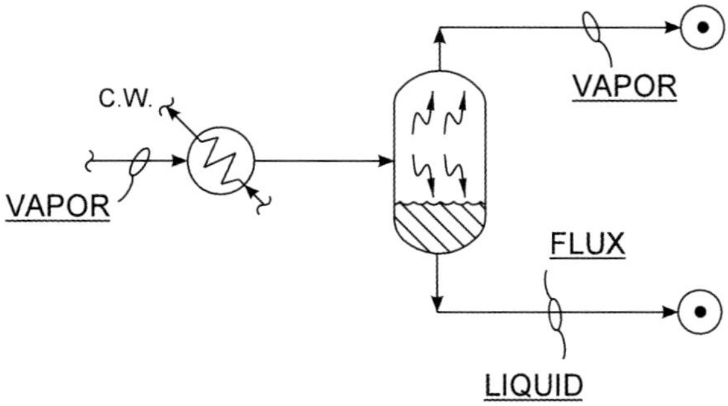

Figure 2.1. Partial condenser

Instead of condensing all of the vapor leaving the still, only part of the vapor would be condensed. Some of the alcohol vapor and some of the water vapor would condense together. However, as the water vapor has a higher boiling point temperature than the more volatile alcohol, the water vapors would condense to a greater extent than the more volatile, or lighter, alcohol vapors.

The relative rate of condensation of the water vapor, as compared to the alcohol vapor, depends on the relative volatility of the water and alcohol (i.e., the ethanol produced from the fermentation of the sugar in the grapes).

Relative Volatility of the Key Components

The definition of the relative volatility of the two components we are trying to separate is explained below:

- The vapor pressure of the lighter component (alcohol).
  ---Divided By---
- The vapor pressure of the heavier component (water).

Let's assume that the vapor pressure of ethanol at 200°F is 20 psia, and that the vapor pressure of the water at 200°F is 10 psia. Then, their relative volatility would be:

$$(20 \text{ psia}) \div (10 \text{ psia}) = 2.0$$

Effect of Concentration

The relative rate of the condensation of water vapor and ethanol also depends on the concentration of water and alcohol in the vapor flowing into the condenser. By concentration, I mean mole % of water, versus the mole % of alcohol. Not their weight percent.

If the mole % of alcohol in the vapor leaving the still was 60%, and the mole percent of water leaving the still was 40%, then the mole % of the first drops of liquid leaving the condenser would be much higher in water. If we collect and discard those first few drops of water-rich liquid into a separate bottle (as shown in Figure 2.1), then we could greatly increase the concentration of alcohol in the remaining vapor stream.

We could do this partial condensation step of the vapors that are leaving our still. But then what would we do with the rest of the uncondensed vapors (Figure 2.1)? The answer to this question forms the basis for all "Distillation," as discussed in detail in the following chapter.

Progression of Distillation

Let us consider the progression of mankind on our planet. Humans advanced from a primitive state in our civilization by discovering:

- The wheel
- Extracting metals from rocks (copper, then iron)
- Starting fires by rubbing dry twigs together
- Befriending baby wolves to become pet dogs

Along with these technical developments was the invention of the "Partial Condenser."

Fermentation of sugar to ethanol occurs naturally. Drinking the fermented mixture was not attractive to primitive mankind. Also, at about 17% alcohol, the fermented sugars become toxic to yeast.

If the fermented liquid was partially vaporized and then totally condensed, at best, the alcohol content of the condensate would be only a maximum of 30%. But if only half of the vapors were partially condensed, the remaining vapors could contain 50% alcohol. Thus, the origin of the "Partial Condenser," which is the basis of modern distillation technology.

The chapters that follow elaborate in detail the complexity that resulted when mankind adopted this principle that forms the basis for our very complex, modern distillation and refinery technology.

The technology I have described in this chapter is still used in 99% of the world's distillation facilities. That is, by people producing whisky and rye in their backyards or garages. To increase the concentration of alcohol in their whisky, they will add additional stages of condensation and generate more reflux. Or even resort to constructing multi-stage distillation towers, which became quite the fashion in the 1920s in Chicago and New Orleans.

Summary

The basis for the separation of components depends on their "Relative Volatility," which is a ratio of the pure components' vapor pressures. The lower the distillation tower's operating pressure, the larger is their relative volatility. A bigger relative volatility allows us to improve the separation of two components (i.e., water and alcohol) with less reflux and a lower reboiler duty, which will save energy used in the reboiler.

Addition of Sugar to Increase Alcohol Content

Adding more sugar to the fermenting mixture will increase its alcohol content. However, at about 16% alcohol concentration, the alcohol becomes toxic to the yeast that is fermenting the sugar in the bottle into alcohol, thus creating a limit on producing brandy with a high percent of alcohol, without resorting to the distillation technology described in this chapter.

Azeotropes

Perhaps you have heard of the term "Azeotropes." Certain components, such as ethanol and water, form a mixture that cannot be separated by ordinary distillation technologies, at certain concentrations. Typically, a third component has to be introduced into the distillation tower to "Break the Azeotrope."

In my work in separating hydrocarbons, azeotropes are not encountered, at least not in petroleum refinery applications. Therefore, I have omitted the various technology countermeasures used to overcome the formation of azeotropic mixtures, other than to note that all azeotropes that I have read about can be overcome with a third component used to "Break the Azeotrope," which is then recovered and recycled back to the distillation tower.

3

THE DISTILLATION TOWER TRAY

The heart of a distillation tower is the tray (Figure 3.1). A tray mixes the up-flowing vapors with the descending liquid. The liquid flow is called the reflux. The rising vapors are called stripping vapors.

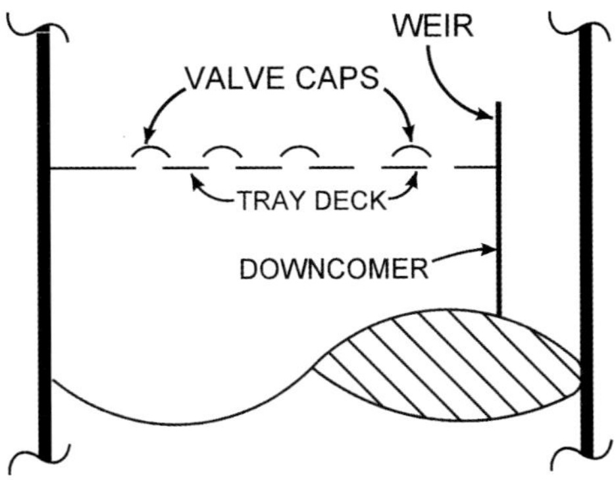

Figure 3.1. Distillation tower tray

The more liquid reflux flow, and the greater the stripping vapor flow, the more effectively the tray works. But only up to the point when the tray becomes overloaded. This is called "Flooding," which is the main factor that reduces a distillation tower's separation efficiency.

The second main factor that diminishes tray efficiency is due to tray deck dumping, or leaking, or weeping. That is, the liquid leaks through a low point of the tray deck (Figure 3.1). The vapor flow rises through the higher points of the tray deck. The result is a reduced vapor-liquid contacting efficiency. The main cause of this all too common problem is that the tray deck is not level because of:

- Careless tray installation
- Upset trays during operation, due to a surge in the vapor flow
- The tray deck thickness, which is too thin for the diameter of the distillation tower (i.e., 16-gauge is bad; 10-gauge is good)

For smaller-diameter towers (less than 4 or 5 feet), the tray deck levelness is a minor problem. For larger-diameter towers (6 to 40 feet), the tray deck levelness can often be a major problem.

Bubble Cap Trays

In the earlier days of the process industry (prior to 1960), the problem that I have described above did not exist. Prior to the 1960s, the industry standard was the "Bubble Cap Tray," which could never leak, because the vapor flowed up through a little chimney inside the bubble cap. This chimney was a bit higher than the top edge of the weir. But bubble cap trays are now used only rarely, at least in the refining industry. And they are only rarely encountered in distillation towers in any plant built since the 1960s.

The negative aspect of bubble cap trays that led to their almost total demise in distillation technology had nothing to do with their capacity or their cost. The real problem was in their maintenance. In order to properly clean a bubble cap tray, each cap has to be:

1. Removed by unscrewing the cap's hold-down nut.
2. Cleaned separately by hand.
3. Replaced by screwing back the hold-down nut.

That's a long task for each cap. If the tray has hundreds of caps, and there are many trays in the distillation tower, that equates to a full week's worth of time for a five-man maintenance crew. And that, in my experience, was the real cause of the demise of the archaic bubble cap tray.

Tray Flooding

If a distillation tower fractionation efficiency is poor due to flooding, then fractionation efficiency will get worse, as the reflux rate and reboiler duty increase. Also, the pressure difference between the top of the tower and the bottom of the tower will gradually increase over a period of perhaps 30 minutes to an hour or two.

More reflux, and more reboiler duty, should improve the fractionation efficiency between the lighter overhead product and the heavier bottoms product. But not if the tray fractionation efficiency is already compromised, due to tray flooding.

Note that there is a complication in what I have just stated. That is, if the flooding is confined to the top one or two trays, then the pressure difference between the top and bottom trays of the distillation tower will not increase noticeably, as only the upper few trays flood.

Tray Dumping or Weeping

If a distillation tower fractionation efficiency is poor because of tray deck weeping, or leaking, or dumping, then raising the reflux rate, and also the reboiler duty, will have a very substantial positive effect on improving the fractionation. The larger the diameter of the distillation tower, the more prone the trays will be to leaking, due to the tray-deck-level problems.

A normal pressure drop for a tray with 24" to 30" tray spacing is about 0.15 psi to 0.25 psi. A much LOWER pressure drop indicates loss of the fractionation efficiency, due to tray-deck dumping. A much LARGER pressure drop indicates loss of fractionation efficiency, due to tray-deck flooding.

Origin of the Reflux Flow

As I have explained above, the observations regarding "Flooding vs. Dumping" are made by altering both the reflux rate and the reboiler duty (i.e., the heat input) together. You cannot simply increase the distillation tower's reflux rate independently of the tower's reboiler duty, or its heat input.

But why?

Because the incremental reflux flow comes from the reboiler. If the control room panel board operator only increases the reflux rate, without also proportionally increasing the reboiler heat input (i.e., duty), then within 15 to 30 minutes, the reflux drum will go empty. And next, the reflux pump will begin to cavitate. It will eventually blow its mechanical seal.

This is not something I have read about in a book. It has happened to me many times in my long career as a Process Engineer.

Sampling vs. Time

The effect on the fractionation efficiency of altering the reflux rate and reboiler duty together is best observed by taking samples and then submitting the samples to the plant's laboratory for composition analysis. Either Gas Chrome (for lighter hydrocarbons) or ASTM D-1160, for residual components.

However, anyone who has ever worked with me knows that I am a very impatient person. I cannot wait for a set of samples to be taken, transported to the lab, have the samples run, and then to eventually receive the results. Typically, this may require several hours, or half a shift.

My method requires 15 to 20 minutes to observe the effect on the fractionation efficiency, due to changes in the reflux ratio and/or the reboiler duty. Just compare the "Temperature Difference" between the top and bottom of your distillation tower.

- If this temperature difference is getting larger, then the fractionation efficiency is improving, which is good!
- If this temperature difference is getting smaller, then the efficiency is getting worse, which is bad!

Optimizing Tower Pressure

I also use the above technique to optimize a distillation tower's operating pressure. I will change a tower's pressure by a few percent (i.e., 2 or 3 psi). And then, over the next 20 or 30 minutes, without altering the reflux rate or the reboiler duty, observe what happens to the distillation tower's "Differential Temperature." If the differential temperature increases, then I have improved fractionation efficiency, which is good. If the differential temperature decreases, then I have degraded fractionation efficiency, which is bad.

My Lesson

Don't sit in your office, attending important meetings with other important people. Go out into the plant and optimize your distillation tower's performance. The key to improved distillation tower performance is direct field observations, combined with using the technical principles that I have enumerated in the following chapters of this book. Regardless of whether you're an operator or an engineer, this advice will help you advance your career in process operations.

Tray Deck Thickness vs. Pressure Drop

For perforated distillation trays, there is a definite correlation between the tray deck thickness and the pressure drop of the vapor flowing through the tray deck's perforations. Somewhat surprisingly, the thinner the tray deck, the larger the pressure drop of the vapor flowing through the tray deck holes!

Very approximately, for a perforated tray deck with a 10-gauge tray deck (quite thick), compared to a 16-gauge tray deck (quite thin), the pressure drop of the vapor passing through the tray deck will be almost twice as high for the thinner tray deck compared to that of the thicker tray deck, for the same flowing, vapor hole velocity.

Summary

The separation efficiency of a distillation tower tray is mainly a result of two factors:

1. Minimizing tray deck weeping, or leaking, or dumping
2. Tray flooding, or entrainment of liquid from a lower tray to the tray above

If the tray vapor load is very low, tray efficiency is compromised due to excessive tray deck weeping or leaking. If the tray load is excessive, then the tray efficiency is compromised due to entrainment of droplets of liquid, blown up onto the tray deck above. A proper tray design seeks to minimize both problems.

4

VAPOR PRESSURES AND RELATIVE VOLATILITY

To really comprehend distillation technology, you will have to understand two very fundamental concepts about all liquids:

1. Their vapor pressure (psia or BARA)
2. Their relative volatility (psia or BARA)

The concept of vapor pressure is something we have all observed from childhood. That is, if I leave a glass of water next to my bed at night, when I wake up in the morning, the water will have partially evaporated. It is the vapor pressure of the water that promotes this vaporization.

If it's been really warm in my bedroom, the water will have evaporated much faster from the glass. If the ambient temperature has increased from 20°C to 40°C, the vapor pressure of the water will have approximately doubled, and the water will have evaporated from my glass twice as fast. The reason for this is that the vapor pressure of the water has doubled by heating the water from 20°C to 40°C. To quantify this effect exactly, you should consult your steam tables or your Mollier diagram.

Relative Volatility

When I was a child, I once became rather ill with the flu. I developed a high fever of 103°F. My mother decided to bring down my body temperature with an alcohol rub (isopropanol). Carelessly, she left a bottle of the isopropanol open by my bedside in the evening. The next morning, I noted that the level of alcohol in the open bottle had fallen far faster than the decline of the level in an adjacent glass of water.

Apparently, I thought, the vapor pressure of the isopropanol (i.e., alcohol) was quite a bit higher than the vapor pressure of the water, at the same temperature. The ratio of the alcohol and water vapor pressures is called their "Relative Volatility":

Vapor Pressure of Alcohol (psia) ÷ Vapor Pressure of Water (psia) = Relative Volatility

Vapor Pressure of Pure Hydrocarbons

Our vapor pressure charts for pure hydrocarbons (Figure 4.1) are a plot of temperature (°Rankine or °Kelvin) vs. the pressure exerted by the pure hydrocarbons. These vapor pressures will rapidly increase, as the liquid in a vessel gets hotter. The ratio of these vapor pressures (for example, propane and normal butane, or C_3 & N-C_4) is the basis for all distillation. That is, the distillation of two hydrocarbons depends on their relative volatility.

The larger their relative volatility, the easier it is to separate two components, such as propane and normal butane. However, it is the nature of all pure hydrocarbons for their difference in vapor pressure to diminish as they become hotter. That is, the cooler a distillation tower operates, the greater the relative volatility between propane and normal butane (Figure 4.1) and all other hydrocarbons.

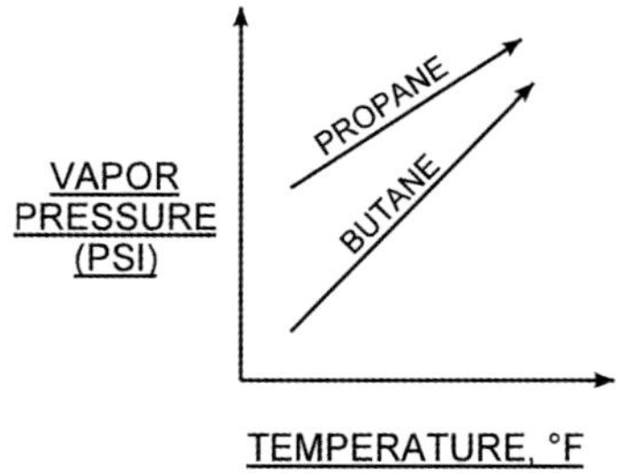

Figure 4.1. Vapor pressure chart

Distillation Tower Operating Pressure

Does this mean that it's easier to separate:

- Propane from butane
- Jet fuel from diesel oil
- Gas oil from asphalt
- Alcohol from water

as the distillation tower's operating pressure and temperature are reduced? Absolutely, YES!

Does this mean that we should operate and design distillation towers at as low an operating pressure as possible? Maybe yes, but maybe no. It's a complicated question, which is at the very heart of distillation technology and is discussed in detail in this text.

Optimizing a Distillation Tower's Operating Pressure

There are two big problems in the reduction of the operating pressure of a distillation tower, with the objective of their enhanced relative volatility and fractionation efficiency. As the tower operating pressure is reduced, the volume of the vapor flow increases. To go from 10 BARA (absolute barometric pressure) to 5 BARA (or from 100 psia to 50 psia) will double the volume of the vapor flow through the trays, which may result in trays flooding, and in the loss of fractionation efficiency.

Also, it's more difficult to condense vapors at a lower pressure. To condense gasoline at 20 psia vs. 40 psia may require that the condenser's cooling water supply temperature be reduced from 90°F to 70°F. Often, that is not possible at most locations.

Energy Efficiency

Reduction in a distillation tower's operating and design pressure will also lower the tower's bottom temperature. And, to some extent, also reduce the reboiler duty, which will save energy. Also, perhaps the reboiler's steam supply pressure may be reduced from 100 psig steam to 50 psig, which also will save some energy.

The enhanced relative volatility at a lower pressure will permit a lower reflux ratio. Since the reflux comes from the reboiler, this will also reduce the reboiler heat load, and also save energy.

All of which is fine, but not if this causes the trays to flood, or if we then run out of condenser capacity in times of warmer weather.

Top Tray Flooding

Unfortunately, there is an added complication relating to an increase in the tower differential temperature used as an indication of improved fractionation efficiency: "Top Tray Flooding."

Often, the tower top tray is subject to fouling from corrosion products generated from the overhead condenser. The fouling deposits cause a high vapor DP across the top tray, which causes the cold reflux to flood over into the vapor line. This reduces the temperature of the vapor leaving the top of the distillation tower, but, unfortunately, does not indicate any improved fractionation efficiency.

Crude Oil Distillation

I have spent a large portion of my career as a chemical engineer and as a Refinery Troubleshooter trying to reduce the operating pressures of crude distillation towers, to enhance their product recovery of kerosene and diesel oil from the crude oil residue, to enhance fractionation between lower value naphtha and higher value kerosene (i.e., jet fuel). My main problem in achieving these objectives is tray deck flooding. If I reduce the crude tower pressure from 3 BARG

(4 BARA) to 2 BARG (3 BARA), then my vapor flow through the tray decks will increase by:

$$(4) \div (3) = 1.33$$

Or by 33%, which only too often will result in a loss of fractionation efficiency, due to flooding, which I cannot tolerate due to an excessive naphtha end-point for gasoline blending.

Distillation Efficiency & Relative Volatility

The vapor pressure of all liquids increases as they get hotter (Figure 4.1). For example, the vapor pressures of both water and alcohol increase as the wine they coexist in is heated to make brandy. This ratio of their vapor pressures is called their relative volatility:

$$\text{Vapor Pressure Alcohol} \div \text{Vapor Pressure Water} = \text{Relative Volatility}$$

As the pressure in the still goes DOWN, which reduces the temperature at which wine evaporates, the ratio of:

$$\text{Lighter components} \div \text{Heavier components} = \text{Increases}$$

This is true for all hydrocarbons, as well as for water and for alcohol. As we reduce their boiling point temperature, we also reduce the temperature that they will boil at. But their difference in boiling points also increases, as they are boiled at a progressively lower temperature and a reduced pressure. That is, lower pressures increase the relative volatility of all hydrocarbons.

Ease of Separation

This means that the lower the operating pressure, the easier it is to make the split between:

- Propane and butane
- Naphtha and jet fuel
- Kerosene and diesel
- Diesel and gas oil

Then why not always run our distillation towers at a low pressure all the time? There are two problems with reducing a distillation tower's pressure:

- Problem #1: The lower the pressure, the more difficult it is to condense the overhead vapor product. This is not a big problem with condensing gasoline that is distilled out of crude oil, but is a much bigger challenge if I'm trying to condense propane that is distilled from natural gas liquids.

- Problem #2: The lower the distillation tower pressure, the bigger the volume of the vapor flow. If a distillation tower pressure is reduced from 45 psig to 15 psig, the volume of vapor flow will DOUBLE:

$$45 \text{ psig} + 15 \text{ psi} = 60 \text{ psia}$$
$$15 \text{ psig} + 15 \text{ psi} = 30 \text{ psia}$$
$$60 \text{ psia} \div 30 \text{ psia} = 2$$

(15 psi is atmospheric pressure.)

The doubling in the volume of vapor, as the distillation tower pressure drops, greatly increases the required diameter of a distillation tower, by a factor of about 40%.

Reducing Tower Pressure: Benefits and Limits

I ran a test in 1982 on a propane-butane fractionator. I was working with my friend Dee, and we were trying to minimize the butane bottoms product. But, at the same time, I had to keep to the spec of not more than 2% butane in the LPG propane, that is, the LPG propane overhead product.

It was a cold evening, so I had plenty of condenser capacity, even at the lower tower overhead pressure. The potential problem was loss of tray fractionation efficiency, due to entrainment of the liquid from the tray below to the tray above. The entrained droplets of liquid would blow butane-rich droplets from the lower tray onto the upper trays, which defeats the purpose of the distillation tower.

"Dee," I explained, "reduction in distillation tower pressure to increase the RELATIVE VOLATILITY of the key components (i.e., propane and butane) is a good idea, but not if the resulting entrainment causes a reduction in "Fractionation Tray Efficiency."

"Norm, the lab is closed. How can we tell whether a reduction in tower operating pressure is making fractionation between propane vs. butane better or worse?"

"Dee, let's try increasing the reflux rate by 15% to start with," I suggested.

"Norm, we'll then have to also increase the tower reboiler duty by the same 15% to avoid loss of liquid level in the reflux drum. But without lab samples, how can we tell if this is helping the split between propane and butane, or making the separation efficiency worse due to entrainment?" Dee asked.

"Dee, we'll watch the tower DT (DT = Tower Bottom Temperature minus the Tower Top Temperature). If DT goes up as we increase the top reflux rate, then the tower fractionation efficiency is getting better, due to enhanced relative volatility.

If DT goes down, then tower fractionation efficiency is getting worse, due to entrainment (also called "Flooding")."

Summary

Using the changes in a distillation tower's temperature profile is my normal method to observe the effects of changing the reflux rate, the reboiler duty, and the tower operating pressure. It's best to wait 15 to 30 minutes for a tower to line-out after making a process change, before drawing any conclusions as to whether fractionation has gotten better or worse.

For me, I use the concept of optimizing tower pressure to minimize the reflux and reboiler duties as an energy-savings strategy, to save reboiler steam consumption. If the fractionator is limited by condenser capacity, this is not a viable option. If, on the other hand, the tower is limited by reboiler capacity, this is a simple way to enhance the reboiler capacity.

Observed tray efficiency vs. design tray efficiency can be compared as follows:

1. Sample the overhead and the bottoms products, and the feed.
2. Measure the reflux rate, the reboiler duty, feed temperature, and pressures (Top – Feed Tray – Bottoms).
3. Run computer simulation, with an assumed number of theoretical stages, until the computer-generated product compositions match the sample results.
4. Divide the assumed theoretical stages for the stripping trays, and for the rectification trays, by the actual number of trays, for each of the two sections (i.e., trays above the feed point, and for the trays below the feed point).
5. Calculate tray pressure drops using published correlations for the types of trays used in the distillation tower (I use the Sulzer engineering correlations).
6. Compare the calculated pressure drops in each of the two sections to the measured pressure drops.
7. If tray efficiency is less than 71%, and the observed pressure drops are low, then the trays are dumping, most likely because the tray decks are out-of-level, or are suffering from a low vapor flow rate.
8. If tray efficiency is less than 70%, and the observed pressure drops are high, then the trays are flooding, most likely because the trays are fouled, or are suffering from an excessive vapor flow rate, or a combination of both factors.
9. Regarding pressure and temperature field surveys, I will do this myself, to be sure that all the plant data is correct and, also, that the data is obtained at steady-state operating conditions, especially that the tower top pressure is maintained constant.

5

THE MEANING OF A THEORETICAL SEPARATION STAGE

A tray in a distillation tower completely mixes the liquid from the tray above and the vapor from the tray below. It then separates both vapor and the liquid phases completely (i.e., no entrained liquid in the vapor). This is called a perfect tray, or a theoretical separation stage, or a theoretical plate, or a tray with 100% efficiency, or an ideal stage.

One characteristic of a perfect fractionation tray is that the vapor and the liquid leaving the tray are both at the same temperature. Also, the vapor has no entrained droplets of liquid. And the liquid leaving the tray is free of froth, or foam, or bubbles, from the vapor phase.

Temperature Profile Refines Tray Efficiency

Assume that I have a section of a tower with the temperature profile as shown in Figure 5.1. This is a sketch of the top section of a refinery crude distillation tower that is fractionating between kerosene (used to make jet fuel) and naphtha (used to produce gasoline or a variety of industrial solvents).

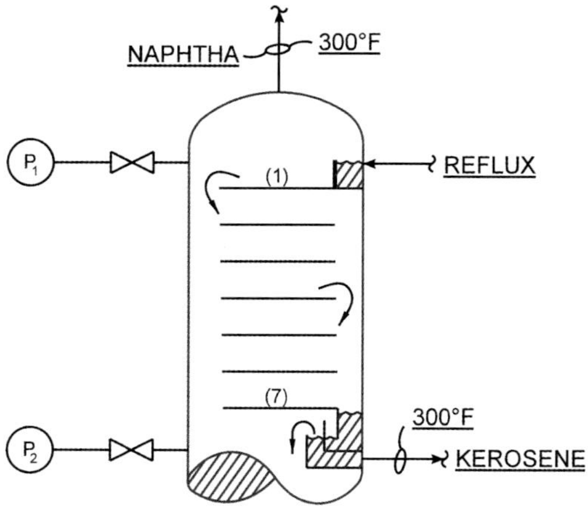

Figure 5.1. Example of a section of a refinery crude distillation tower fractionating between kerosene and naphtha

Between the kerosene side draw-off and the tower top, there are seven valve, type trays shown. The vapor temperature from the top tray (300°F) and the liquid draw-off temperature from Tray #7 are both 300°F. This is bad! This indicates that the seven trays are developing a fractionation efficiency equal to a single theoretical separation stage. Or, just a single perfect tray, with 100% fractionation efficiency. Therefore, we would calculate the observed tray efficiency as follows:

$$(100\%) \div 7 \text{ trays} = 15\%$$

This represents very poor tray fractionation efficiency. Good would be 65% to 80%. Poor would be 40% to 50%.

This very poor fractionation efficiency may be due to severe entrainment of the liquid from the tray below due to flooding. Or it may be due to excessive tray deck dumping, or weeping. To differentiate between these two very common causes of loss of tray fractionation efficiency, I will next run a pressure drop survey between trays one and seven, as shown in Figure 5.1.

Pressure Profile

If the pressure difference between P_2 and P_1 is low (i.e., less than 0.5 psi for the seven trays shown), then the fractionation efficiency between the naphtha and kerosene is bad, because the trays are dumping excessive amounts of liquid through the tray deck. This is called loss of fractionation efficiency, due to tray deck weeping, or dumping, or leaking. This is a very common problem that I see repeatedly.

If the pressure difference between P_2 and P_1 is high (more than 2.0 psi), then the fractionation between naphtha and kerosene is bad, because the trays have excessive entrainment of droplets of liquid, from the tray below, blowing up to the tray above. Usually, this is due to excessive vapor velocity between the trays. Or, perhaps, excessive liquid back-up from the downcomers, which may be too small. Or the downcomer clearance is undersized, and the liquid level in the downcomers is backing up. That is, the trays are flooding.

The usual cause of loss of fractionation efficiency, due to tray deck dumping, is that the tray decks are more than one inch out-of-level. Out-of-levelness should not exceed 0.5". The usual cause of tray deck flooding, and/or downcomer back-up, is that the tray decks are fouled.

Using Temperature to Evaluate Tray Efficiency

Again referring to Figure 5.1, if the vapor leaving the top tray (300°F) is the same as the liquid leaving tray #7 (also 300°F), then these top seven trays are performing like a single theoretical separation tray. Therefore, the observed tray efficiency is only:

$$(100\%) \div 7 = 14\%,$$

which is terrible. A usual design tray efficiency used in computerized simulation is 65% to 80%. A very conservative tray efficiency to use in a new design is 50%. A very optimistic tray efficiency to be used in a new design is 90%. I use 70%.

Direct Field Observations

For me, as the world's most experienced Refinery Process Engineer (I started work on November 19, 1965, at 9:30 am)—i.e., 61 years ago—I find my understanding as to how distillation towers work is greatly aided by direct field observations, using:

1. My infrared surface temperature gun
2. My 0-30 psi pressure gauge, for towers operating at less than two BAR (30 psig)
3. My digital battery-powered vacuum pressure gauge, for towers operating at ten to 150 mm of Hg)

Very often, in the course of making direct field measurements, I have discovered the reason for the poor fractionation efficiency of the distillation tower, or of a sour water stripper, or of the refinery vacuum tower, by accident.

Common Causes of Distillation Tower Premature Flooding Due to Mis-assembly or Design Errors

1. The downcomer bracing brackets have not been installed, or not tightened sufficiently.
2. The downcomer clearance is too small to the tray below.
3. The downcomer clearance is too large to the tray below. Thus, the downcomer seal has been lost.
4. The weir height is too small, and thus the downcomer seal has been lost.
5. Fixed valve cap tray decks have been installed backwards, which is a very common problem.
6. The bottom tray downcomer seal pan does not have a sufficient clearance to the bottom downcomer.
7. Hats on the top of chimney trays are installed too close to the top of the chimneys.
8. A vortex breaker, on a side draw-off product, is restricting the draw-off nozzle's open area. A common problem.
9. Excessively long downcomers have been accidentally supplied by the tray vendor.
10. Lack of a drain hole in bottom downcomer seal pan.

Summary

How close a distillation tower tray is operating to a 100% efficiency of a theoretical stage is essential to understanding how well a distillation tower is operating. This is done with both a vertical temperature survey, and by a sample analysis of the overhead and the bottoms products, as compared to a computer-generated simulation.

To discriminate between tray deck flooding vs. tray deck dumping, a field pressure drop survey is the next step. It is best to perform this pressure drop survey oneself. I typically perform this critical survey myself, using new pressure gauges I have personally installed.

Tray mis-assembly is the leading cause of poor distillation tray efficiency. I have solved more distillation problems by personal tray field inspection after their installation than by running computer process simulations. The key to solving distillation problems is not in your office, but outside in the plant.

The Patent Still

The Patent Still was developed in the 1880s. It was used to produce various sorts of alcoholic beverages. I saw such a still in 1990 in an apple orchard in Somerset, England. As I recall, it appeared to have a cast iron shell. It was roughly 18" to 24" I.D., and had about 8 cast-iron bubble cap trays.

For a "Reboiler," it burned firewood. A local stream provided cooling water for the condenser. It was in excellent condition and appeared to have been used in the previous year or two.

All early distillation towers came with the trays—which were always bubble cap trays—installed as an integral part of the distillation tower itself. I often wonder whether this was not a superior form of technology, compared to current practices.

6

DISTILLATION PROCESS CORRELATIONS AND CALCULATIONS

In 1966, I was the first person in the world to design a complex refinery fractionator, using a computer program ("B" coker, American Oil, Texas City) that was actually built. It had side-stream steam strippers (for light and heavy coker gas oil), pumparounds (middle and lower), and a two-stage overhead condenser (air-cooled).

I used a computer tray-to-tray process simulation model that I wrote myself for this purpose. (Incidentally, this tower is still in service next to Highway 1764 in Texas City.)

But, as I progressed in my career, I have fallen back, in my day-to-day work on multi-component fractionator troubleshooting, to using the following manual calculation correlations.

General Correlations

Note: always use absolute pressure and temperatures, not psig or °F, or °C or BAR. I use PSIA, or °R (or °K, BARA).

Gas Volume

$$\text{Volume of gas (cubic feet)} = (379) \bullet (N_1) \bullet (14.7 \div P_1) \bullet (T_1 \div 520)$$

Where:

N_1 = (lbs. of vapor) ÷ (M.W.)
(M.W. = Gas Molecular Weight)

P_1 = Pressure, psig + 14.7 (or BARG + 1.0)

T_1 = Temperature, °F + 459 (or °C + 379)

Pressure Drop

$$\text{Delta P} = K \bullet \text{Density} \bullet (V)^2$$

Where:

K = Orifice coefficient
Density of flowing fluid
$(V)^2$ = Velocity, squared

For orifice coefficients (K) through a nozzle restriction or orifice plate:

K = 0.15 (theoretical coefficient)

K = 0.34 (typical for nozzles)

K = 0.5 (orifice plates for flow indications) (when working with psia)

Vapor Pressure

Let's say I have a bottle of pure propane. On top of the bottle, I place a pressure gauge and then heat the bottle to 100°F. The pressure gauge reads 285 psig. If I now add 15 psi (atmospheric pressure):

$$285 + 15 = 300 \text{ psia}$$

I have obtained the vapor pressure of propane at 100°F. If working with metric units, I would add 1.0 to convert from gauge to an absolute pressure.

The Ideal Gas Law

$$PV - \text{proportional} - NT$$

Where:

T is expressed degrees Kelvin or in degrees Rankin (not °F or °C)

P is expressed in psia (not psig), or in absolute barometric pressure

V is expressed in cubic ft. or cubic meters

N is the number of moles of vapor

Downside of Reliance on Computer Calculation and Simulations

Very often, as I perform my calculations manually, design opportunities to improve an existing distillation tower performance occur to me, that I would not have thought of if I had been presented with a computer-simulated tray-to-tray output. Manual calculations, for me, simulate innovative thinking about process problems, and possible cost-effective solutions. As example of this problem

I will discuss is heat extraction in a distillation tower pumparound. With the pumparound circulation rate increased, the objective is to increase heat extraction from the lower section of the distillation tower.

Pumparound Heat Extraction Duty

Let's assume I have increased the pumparound circulation rate by 20%, which is a typical operational change. If we use a computer simulation to calculate the new, higher pumparound heat extraction rate from the distillation tower, will the pumparound heat extraction increase also by 20%?

Absolutely not! But why?

Because, the increased pumparound circulation rate will extract some increment of heat from the tower, which will:

1. Condense an increment of lighter vapor components into the pumparound draw-off flow.
2. These lighter components will then reduce the pumparound draw-off temperature.
3. This reduced pumparound draw-off temperature will, unfortunately, reduce the heat transfer temperature driving force (LMTD) in the pumparound heat exchanger.

Not a problem! Your computer simulation will correct for this reduced draw-off temperature and reduce the pumparound heat extraction, based upon the designated pumparound return temperature you have specified.

But here's the problem. The actual pumparound return temperature will increase as you increase the pumparound circulation rate.

But why?

Because the pumparound temperature driving force on the hot-end has gone down. And the cold-end (i.e., the pumparound cold-end return temperature) has gone up, because you have increased the pumparound circulation rate!

Benefit of Manual Calculations

As the world's most experienced refinery crude distillation tower process design engineer (60+ years), these sorts of problems only occur to me when I supplement a computer simulation output with a manual calculation for certain variables. By all means, let's use our computer simulations as a basis for design. I do. But let's also understand that, sometimes, manual adjustments need to be made to our simulation outputs.

Selection of "Equation of State"

When we run a computer simulation, we must select which "Equation of State" to use. My only method to select the appropriate equation of state is to select an operating unit in a similar service and determine which equation of state represents the observed operation more accurately.

Often, I see engineering contractors use the default equation of state. This means that their design for their distillation towers is based on a guess, which only too often does not coincide with reality.

Summary

In process engineering, we have to understand how the equipment works. Use of manual calculations to supplement, but not to replace, computer simulations is my method to satisfy this basic engineering requirement.

Selection of tray efficiency is, at best, a guess for a new distillation tower. I will typically guess at 70% to 75%, assuming that the tray is installed correctly. Selection of the equation of state to be used in a particular service is always a difficult question, which can only be correctly answered if there is a distillation tower in a similar service that can be modeled and will serve as a prototype for our current Distillation Tower design.

7

MULTI-COMPONENT DISTILLATION

"You know, Norman," Liz said, "I did study distillation technology at the university. Our professor explained distillation using the 'McCabe-Theil Diagram.' So, I really do understand the fundamentals of multi-component distillation."

"What you studied at the university, Liz, was only 'Binary Fractionation Technology.' For example, separation of propane and normal butane in the absence of other components, such as methane, ethane, hydrogen sulfide, normal pentane, hexane, and heptane."

"What's the difference," Liz asked, "if my only interest is to keep the design product specifications of:

- 2% normal butane in the propane overhead product,
- 4% propane in my normal butane bottoms product?"

Effect of Non-Key Components

"Liz, referring to Figure 7.1, let's assume that I have 20% butane in my overhead product, which is a mixture of:

- Methane
- Ethane
- Propane
- Butane

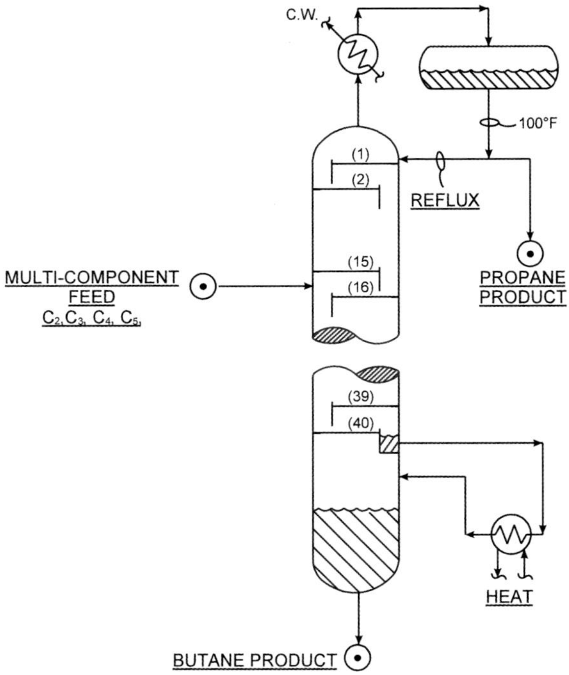

Figure 7.1. Propane-Butane Fractionator

Let's also assume that the reboiler duty is constant, and that the feed temperature and percent vaporization are also constant. That is, the heat flow up the tower is constant."

"Okay, Norm," Liz said, "all you're stating is that the enthalpy and the pounds of vapor leaving the reboiler are constant."

"Correct, Liz! And if I now increase the reflux rate to lower the concentration of normal butane in the propane overhead product, then the tower top temperature will..."

"Go down," Liz stated. "That's obvious, Norm! So, what's your point?"

"My point, Liz, is that if I lower the tower top temperature, so as to reduce the butane concentration of the overhead propane product, then each pound of the vapor leaving the top of the tower contains less heat, because it's cooler. But since the reboiler duty is constant, then the pounds of vapor leaving the top tray of my distillation tower must increase!"

"Yes, that's true, Norm. But where does that extra vapor come from?" Liz astutely asked.

"It comes from the extra reflux," I replied.

"But, Norm," Liz asked, "where does that extra heat to generate this extra reflux come from?"

"Liz, it comes from the reduction in the temperature (i.e., sensible heat content) of the vapor leaving the top tray. As we increase the tower top reflux rate, a portion of this extra reflux is evaporated. The heat of vaporization for this increment of vapor flow, to generate this extra reflux flow, Liz, comes from the reduced sensible heat content of the up-flowing vapors," I explained.

"Norm, could this extra vapor flow, which is generated by the incremental evaporation of the reflux, due to the sensible heat reduction of the up-flowing vapors, cause the upper trays of a multi-component distillation tower top to flood?" Liz quite perceptively asked.

This is a really good question that is not well understood in towers in multi-component services.

Effect of Non-Key Components in Distillation Towers

"Liz, what you have just asked is the leading cause of flooding in refinery fractionators that are in:

- Crude units
- Delayed cokers
- Fluid catalytic crackers
- Wax fractionators
- Vacuum towers
- Lube oil distillation towers
- Visbreakers

This flooding occurs on the upper trays of multi-component fractionators. If I am fractionating between pure components such as propane and butane, the effects I've, described are quite small. But in multi-component distillation, especially in crude and coker fractionators, it is one of the main causes of flooding, and of poor fractionation, between naphtha and kerosene. That is, the premature flooding of a distillation tower's upper few trays.

Optimizing Pumparound Heat Removal

A large percentage of distillation towers are equipped with pumparounds, which are used to remove heat from the center portion of a distillation tower. This will naturally diminish the distillation tower's top reflux rate.

Ordinarily, the reduction in the top reflux will diminish the fractionation efficiency, which is bad. But if the trays above the pumparound have been losing fractionation efficiency due to entrainment or flooding, then the reduced vapor velocity, which results from the increased pumparound heat extraction, will improve the fractionation efficiency. To see if increased pumparound rates are aiding or are hurting the fractionation efficiency, check the differential temperature between the pumparound draw-off tray and the tower top temperature.

- If this differential temperature is increasing, then fractionation is getting better.
- If this differential temperature is decreasing, then fractionation is getting worse.

Water in Reflux

Perhaps the single most common cause of premature flooding in many distillation towers is not the excessive vapor or liquid flows, but having free water in the reflux. This free water forms an aqueous phase on the top tray, which promotes corrosion, fouling, and salt formation.

Very often, I sample the reflux flow, and find free water mixed with the hydrocarbon liquid reflux flow. The most common cause of free water in the distillation tower's reflux is that the reflux nozzle riser pipe (see Figure 7.2), is too short. Its height should be about twice the diameter of the hydrocarbon (i.e., reflux) outlet nozzle, as shown in Figure 7.2 (Nozzle "D").

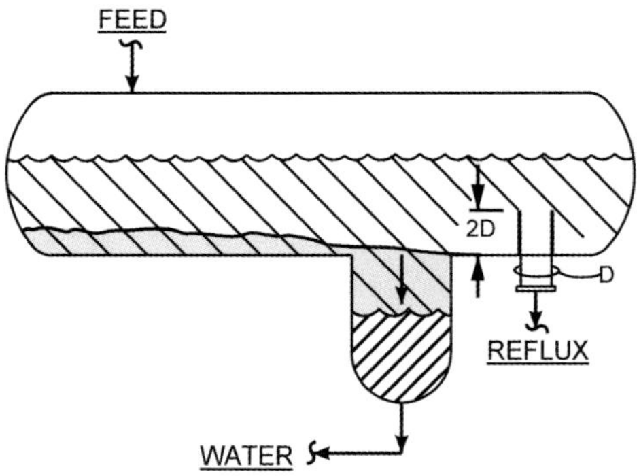

Figure 7.2. Reflux nozzle riser pipe

Summary

"Well, Norm," Liz concluded, "I am certainly very interested in the problem of multi-component fractionation. But what can be done to avoid this top tray flooding?" Liz asked.

"Keep the upper trays DRY to prevent corrosion due to salt formation. Fouling of the upper trays causes loss of tray capacity and flooding."

This is the single biggest problem that we have in all refineries worldwide. Not a month goes by that I don't have a client who experiences this sort of problem.

Non-key components in the distillation tower affect the separation efficiency of the key components, leaving water in the reflux and invalidating all process engineering calculations regarding distillation tower fractionation efficiency. It's a major problem in my work as a Distillation Engineering Consultant. It's best to get a sample of reflux and check visually for any free water.

8

HOW DISTILLATION TRAYS WORK

A tray is a vapor-liquid mixing device. Vapor bubbles up through the tray caps, perforations, or holes. The vapor then mixes with the liquid running horizontally across the tray's deck. If the tray has a 100% efficiency, both the vapor leaving the tray and the liquid from the tray will have the same temperature and be in equilibrium.

If the vapor leaving the tray is hotter than the liquid leaving the tray, then the tray fractionation efficiency is less than 100% (Figure 8.1). A typical good tray efficiency is 70% to 80%. Often, tray efficiency is bad. That is, 0% to 20%! Poor tray efficiency may be due to two sorts of problems: tray deck flooding or dumping. Both are very common problems, mostly due to tray deck fouling, which causes flooding. Or out-of-level trays, which cause dumping.

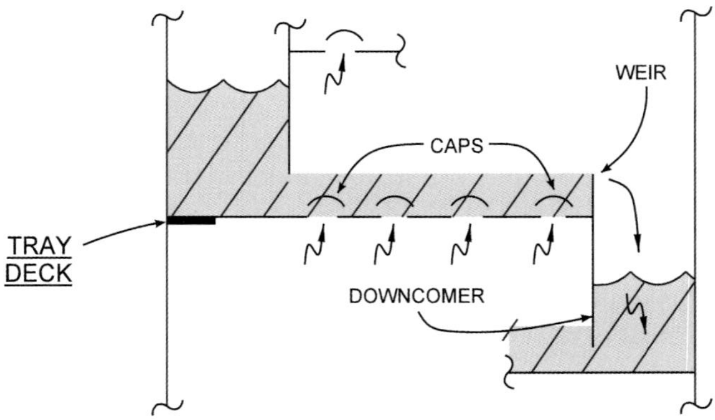

Figure 8.1. Distillation tray

Tray Dumping (Weeping)

The valve caps on a tray cannot stop the liquid on the tray from leaking through the tray deck. It's the pressure drop of the vapor, as it accelerates through the holes on the trays, that prevents the tray from leaking or weeping. If the tray deck is very level (i.e., not more than 0.25" to 0.40" out-of-level), then the tray deck leakage would not compromise tray efficiency too much. But if the tray is badly out-of-level by 1" or more, then most of the liquid may leak through the low points, and not be able to come in good contact with the rising vapors flowing through the higher points on the tray deck.

The result is poor vapor-liquid contacting, and thus low tray fractionation efficiency.

The greater the weight of the liquid on the tray, the greater the tendency for poor tray efficiency, due to dumping. The weight of liquid on the tray is mostly a function of the weir height. That's the ridge at the edge of the tray (Figure 8.1). A typical weir height is 2" to 3". Most common are 2½" weirs.

As the liquid overflows the weir, it forms a crest. This crest height depends on the liquid flow rate and is typically 1" to 1½."

The big problem you will have for trays that lose tray efficiency due to dumping is out-of-level tray decks. Or trays that have too many holes for the vapor flow. Or a combination of both low vapor flows and out-of-level tray decks.

Tray Flooding

Distillation trays may also do a bad job of fractionation due to liquid droplets being entrained from the tray below to the tray above. This entrainment blows the liquid from a lower tray to the tray above. This back-mixing of the liquid reduces the fractionation efficiency of a tray deck.

Entrainment of the droplets of liquid from the lower tray to an upper tray occurs from a combination of two bad factors:

1. Downcomer back – The liquid overflowing a tray's weir (see Figure 8.1) then flows into a downcomer on the side of the tray. There are many factors that can restrict the downflowing liquid flow. These factors are:
    a. The cross-sectional area of the downcomer is too small.
    b. The liquid in the downcomer foams up, because of particulates in the distillation tower's feed, or water mixing with liquid hydrocarbons.
    c. The downcomer clearance, above the tray deck below, is too small (less than two inches).
    d. There could be a hole in the downcomer that permits vapor to flow into the downcomer and causes it to back up and flood.
2. Droplet entrainment – High vapor velocity, combined with high vapor density, can cause droplets of liquid to blow upwards from a lower tray to the tray above. This back-mixing is called "Jet Flood," or "vapor flooding." The smaller the tower's tray spacing (i.e., space between trays), the greater the tendency to cause jet flood due to entrainment.

Most trays are set on a 24" or 30" tray spacing. A small tray spacing is 18." A very large tray spacing is 36."

Increasing a distillation tower reflux rate too high causes excessive vapor velocity, entrainment, and eventually a loss of fractionation efficiency due to jet flood. The flow of reflux originates from the vapor generated in the reboiler, which then condenses in the overhead condenser into reflux.

Why Fractionation Efficiency is Lost

To discriminate between flooding and dumping, both of which may be causing a loss of fractionation efficiency, there are two methods:

- DT = Temperature Difference
- DP = Pressure Drop

A typical pressure drop across a tray is 0.2 psi per tray. If the DP is 0.4 psi to 0.6 psi per tray, you can assume fractionation efficiency is bad because of flooding or entrainment. If the observed pressure drop is less than 0.10 psi per tray, you can assume fractionation efficiency is bad due to tray deck dumping, weeping, or leaking. Of course, these rules also depend on the:

- Tray spacing
- Specific gravity of the liquid
- Foaming tendency
- Tray deck levelness

Regardless, a tower pressure drop survey, using a single, calibrated pressure gauge, is always my starting point in determining the causes of poor fractionation efficiency.

Temperature Difference

My favorite method to discriminate between flooding and dumping of a trayed tower is to:

1. Note the temperature difference between the top and bottom of the tower (DT)
2. Raise the reflux rate and the reboiler duty by 10%
3. Observe what happens to the "DT"

If the "DT" goes down as I raise the reflux rate (and reboiler duty), then the tower is suffering from tray flooding, and I am making the problem worse with more reflux flow. If the "DT" goes up as I raise the reflux rate (and the reboiler duty), then the tower is suffering from tray deck dumping, and I am improving the fractionation efficiency. Which is good!

Of course, I could run the above tests by getting samples of the overhead and bottom products, but it takes too much time to wait for the lab results, and I am always in a big rush. However, time permitting, it is always best to rely on samples and lab results.

Calculating Tray Efficiency

A typical distillation tray efficiency is 65% to 80%. To calculate the actual tray efficiency of a distillation tower, we would proceed as follows:

1. Check the reflux rate and the reboiler duties.
2. Obtain a sample of the bottoms and of the overhead products.
3. Using the flows from step 1, alter the number of theoretical trays in the "Tray-to-Tray" computer simulation for the distillation tower until the computer-generated product split matches the sample results for the products.
4. Divide the number of theoretical trays in the computer simulation by the actual number of trays in the distillation tower.

The result of step 4 is the observed "Distillation Tray Efficiency." If the tower pressure is low, the poor tray efficiency is due to tray deck dumping, or weeping. If the tower pressure drop is high, the poor observed tray efficiency is due to tray deck flooding, or excessive entrainment.

Total Trap-Out Chimney Tray

When drawing off a side product from a distillation tower, it is best practice to use a "Total Trap-Out Chimney Tray" with provision for internal overflow of liquid to the tray below. Using an ordinary fractionation tray for a side draw-off may not allow the desired side draw-off rate to be achieved. The potential problem is that the tray may leak to an extent that there will not be enough flow into the tray's draw-off sump to satisfy the desired side draw-off flow rate.

Of course, the fractionation efficiency of the chimney tray is ZERO, because the chimney tray has no vapor-liquid contacting. But that is the price I have to pay for ensuring that I can maximize my side draw-off product rate, even at a very reduced vapor flow through the draw-off tray.

Summary

The two most common causes of poor fractionation in trayed distillation towers are:

- Flooding, due to fouling on the tray decks, or not enough downcomer clearance,
- Dumping, due to the tray decks being excessively out-of-level, or excessive weir heights.

To minimize these problems, I always use trays with a fixed valve cap. I never use trays with movable valve caps, which can get stuck in an open position. Also,

I always specify tray decks to be constructed from 10-gauge rather than 14-gauge metal. The 10-gauge is quite a bit thicker (about 40%) than the 14-gauge metal that is normally used by tray manufacturers. Thick tray decks are inherently more level, and less prone to bending during their installation.

The industry standard for maximum tray deck "out-of-levelness" is 5/8". This is easy to obtain in a 3 ft. diameter tower, but quite a challenge for the 17'-6" I.D. delayed coker fractionator that I designed for a refinery in California in 2015.

Tray fractionation efficiency is negatively impacted by tray deck:

1. Flooding (entrainment)
2. Weeping (tray deck leaking)

Tray efficiency is most readily calculated from the distillation tower temperature profile. A compressed temperature profile is a positive indication of loss of distillation tower tray fractionation efficiency.

To discriminate between loss of tray fractionation efficiency due to tray flooding, or weeping, a field pressure drop survey you do yourself, is the cheapest, quickest solution.

A radiation scan is a more expensive alternate, but often reveals more precisely the actual location of the problem, causing the loss of tray fractionation efficiency (see Chapter 31 for a complete description of a Radiation Scan Survey). I am a big fan of such radiation scans, but rarely use them because of the time it takes to obtain management approval for the required funds – typically $10,000 - $30,000.

I specify tray decks to be constructed from 10-gauge rather than 14-gauge metal. The 10-gauge metal is quite a bit thicker (about 40%) than the 14-gauge metal that is normally used by tray manufacturers. Thick tray decks are inherently more level, and less prone to bending during their installation.

The industry standard for maximum tray deck "out-of-levelness" is 5/8". This is easy to obtain in a 3 ft. diameter tower, but quite a challenge for the 17'-6" I.D. delayed coker fractionator that I designed for a refinery in California in 2015.

9

DISTILLATION TRAY DESIGN AND VORTEX BREAKER FOULING

A distillation tray is a vapor-liquid mixing device. It mixes the up-flowing vapor with the descending liquid flow (called reflux). The better the mixing, the better the performance of the distillation tower.

If the mixing on the tray is perfect, then the vapor and liquid leaving the tray will have the same temperature, and are said to be in EQUILIBRIUM. Such a tray will have 100% efficiency. However, a typical tray will have an efficiency of 60% to 80%. But if the tray deck is leaking, its efficiency may only be 30% to 40%. If the liquid from a lower tray is entrained to the tray above, then tray efficiency will also be compromised and may be close to zero.

Fractionation Tray Design

As a liquid flows across a tray's deck, the liquid mixes with the vapor flow from the tray below. In order for the vapor and the liquid to have enough time to mix efficiently, a liquid level of several inches is maintained on the tray's deck by a weir, typically 2 ½" high (see Figure 9.1).

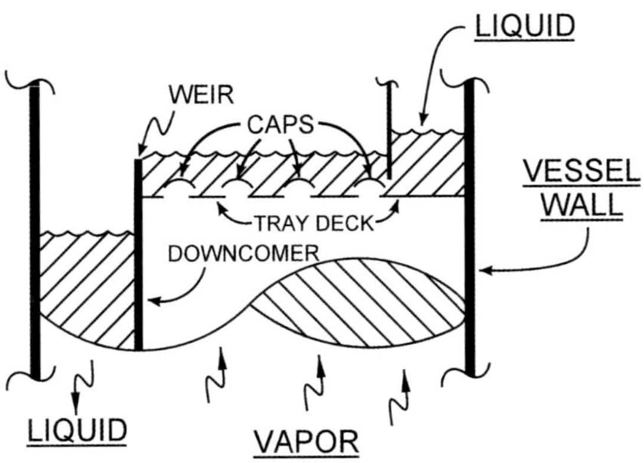

Figure 9.1. Tray components

The liquid overflows this weir, into a downcomer. The downcomer is a vertical plate bolted to vertical strips (which are called downcomer bolting bars). The vessel wall and the vertical downcomer plate form the downcomer. The downcomer then feeds liquid onto the inlet portion of the tray below. This liquid runs out onto this lower tray through a gap between the bottom edge of the downcomer and the tray below (i.e., the downcomer clearance).

This gap is typically 2" or ½" below the weir high. The part of the tray below the downcomer has no perforations. As the liquid escaping from the downcomer flows across the perforated section of the tray, it mixes with up-flowing vapors.

This is the purpose of a distillation tray: to mix the up-flowing vapor with the down-flowing liquid. The more efficient this mixing of vapor and liquid, the greater the "Fractionation Efficiency" of the "Distillation Tray" (Figure 9.1). After the liquid mixes with the vapor leaving the tray deck, the vapor flow rises to the tray deck above (Figure 9.2).

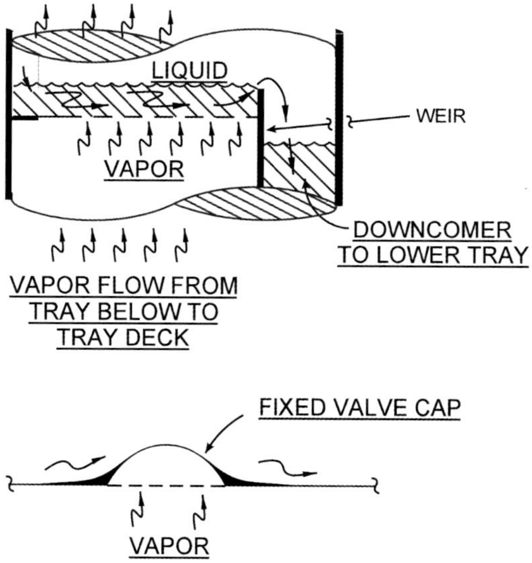

Figure 9.2. Fractionation tray

Tray Deck Perforations

The tray deck is perforated with a varied assortment of holes or valve caps. The simplest sort of tray deck is called a "Sieve Tray." This is a tray with ½" or ¾" holes. Typically, perhaps 10% to 12% of the tray deck area is devoted to perforations, through which the rising vapors flow.

Often, these perforations are partially covered with caps. This is called a valve tray. Most trays have valve caps rather than sieve holes. In older days (prior to 1960), most trays had "Bubble Caps." Bubble caps are only very rarely used today.

Downcomer Design and Flow

The liquid that runs across the tray deck overflows a weir about 2" to 3" high into the downcomer. Typically, the outlet downcomer may encompass 20% of the tray's area at the overflow-side of the tray, plus another 20% of the tray's area at the liquid inlet-side of the tray (Figure 9.1).

Effect of Flow Path Length

The distance between the inlet liquid side of the tray and the outlet weir is called the flow path length. This can be as small as 20" for a tiny 30" I.D. tower. For larger towers, we do not allow the flow path length to be much longer than 5 ft. (60"). A much longer flow path length would create a hydraulic gradient for the liquid flow. That is, the liquid height would diminish across the tray deck. This would distort the vapor flow, and force the vapor to flow preferentially to the far end of the tray, closest to the outlet weir. To keep this flow path length from getting too long, we use two-pass trays (Figure 9.3). These have dual, parallel, liquid pathways.

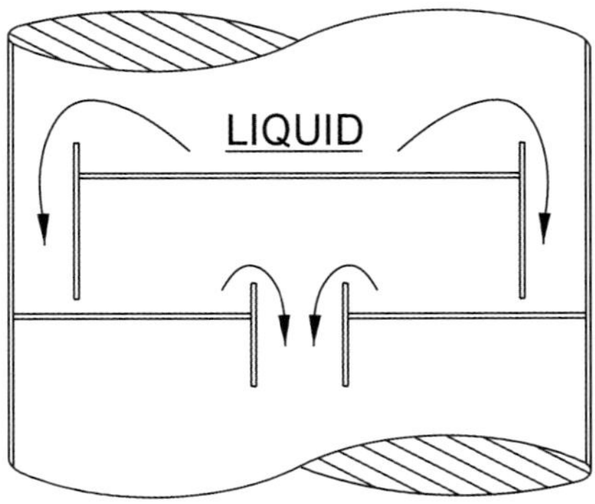

Figure 9.3. Pass tray

Tray Deck Spacing

The industry standard for smaller-diameter towers (3 ft. to 6 ft. I.D.) is to use a tray deck spacing of 24." For larger-diameter towers (8 ft. to 12 ft. I.D.), a tray deck spacing of 30" is typical. For much larger towers, either 30" or even 36" tray spacing is used.

Tray Deck Levelness

The major problem in distillation technology, which is not well understood, is tray deck levelness. The industry standard is maximum out-of-levelness of 0.6," which is easy to achieve in a tower I.D. of 36." But, quite a challenge in the 17 ft. 6" I.D. fractionator I designed in 1991, for a delayed coker in California.

One way to help achieve greater tray deck levelness is to make the tray decks out of thicker metal (10-gauge). But the industry standard is to use thinner metal (i.e., 16-gauge or 14-gauge).

Tray Efficiency

The key to process plant or refinery profitability is good distillation efficiency. The key to distillation efficiency is good tray efficiency. The key to achieving good tray efficiency is to understand how the trays work and to inspect the tray installation during the plant turnaround. Also, it is to review the tray drawing prior to the turnaround, or the purchase of the new trays.

Downcomer Loading

Downcomer loading is based on gallons per minute, per $ft.^2$ of downcomer areas.

For non-foaming services, assuming a 24" tray spacing, for a straight downcomer, I would use a maximum loading of 150 GPM (gallons per minute) per $ft.^2$ of downcomer, top, cross-sectional area. For foaming services, a reasonable downcomer loading, is 90 GPM/ft^2, on the above basis.

Downcomers are very often sloped, so that the bottom open area could be two thirds of the downcomer top area. For 2-pass trays, I maintain the same downcomer areas for both the center and the side downcomers. Increasing tray spacing from 24" to 30" permits a very small increase in the downcomer loading, perhaps by 10% to 15% at most.

Low Weir Loading

A leading cause of poor tray efficiency is low weir loading. Any weir loading of less than 2 to 3 GPM per inch of weir length will result in poor liquid distribution on tray decks and low tray efficiency.

This is best corrected for by reducing the effective weir length by installing extended sections of elevated portions of the weir. Usually, these extended sections are 8" to 12" in height and perhaps 6" to 8" wide. Their effect is to increase the depth of liquid on tray decks by one or two inches. A minimum weir loading of 2 to 3 GPM per inch, between the pickets, is a good criteria. Installation of these sorts of picket weirs is one of my common tray retrofit features.

Draw-Off Nozzle Plugging Due to Fouled Vortex Breakers

Strange to say, this is likely the single biggest problem, at least in distillation towers, in the refinery industry. It is more important than the lack of computer control, or radiation liquid-level indication. If the draw-off nozzle calculated outlet velocity exceeds 4 ft./second, then a vortex breaker is certainly needed.

If the draw-off nozzle calculated outlet velocity is less than 2 ft./second, then a vortex breaker is certainly not needed.

In a new amine absorber-stripper that I designed recently, I oversized all the draw-off nozzles for 2 ft./second, and thus omitted all the vortex breakers. This, I believe, will enhance the on-stream reliability of the new amine system that I designed for Venezuela.

Time will tell! I developed the above nozzle vortex correlations from my bathtub, at my home in New Orleans.

Summary

Detailed distillation tray design is a complex issue, best left in the hands of the two major tray design and fabrication contractors in America. However, like all contractors, they can make mistakes. Or they may reduce the mechanical integrity of their tray designs to cut costs. I always find it necessary to carefully review all tray data sheets and tray drawings (and also details for beds of structured packing, especially the structured packing internal reflux distributors).

In the course of making such reviews, I may identify an area where the mechanical integrity of a tray, or of a packed bed, could be increased, at some small incremental cost. A simple example is using 10-gauge rather than 16-gauge tray decks. Or drilling ½" holes through the tray support rings of a small-diameter tower, to enhance its fractionation efficiency and its capacity.

10

TRAY TYPES – TRAY DECK LEAKAGE

"Norm, I've heard from an older engineer that trays used to be made with a 3" or 4" high chimney, inside a cap?" Liz asked. "Is that right?"

"Yes, Liz. Those were called bubble cap trays. Since the height of the vapor chimneys was higher than the overflow weir (Figure 10.1), such trays could never leak if they were mechanically intact," I responded.

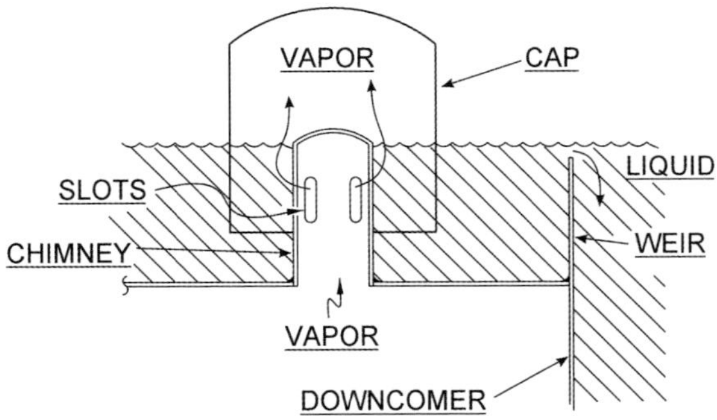

Figure 10.1. Bubble cap trays

"In the early 1960s, tray manufacturers stopped making bubble cap trays and started to make cheaper valve trays (Figure 10.2), which, as you can see from my sketch, could leak."

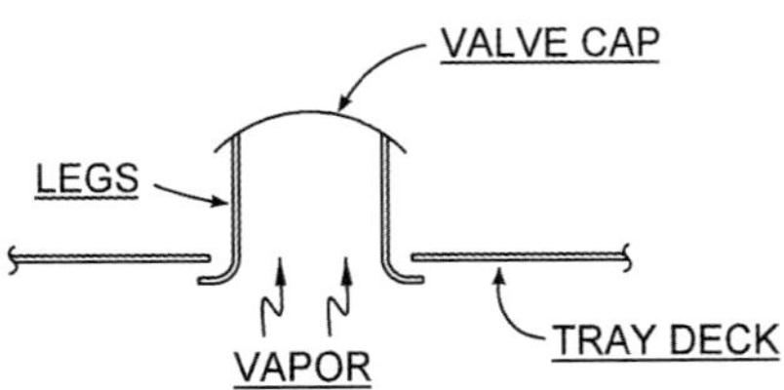

Figure 10.2. Valve cap tray

"But, Norm, without the vapor chimney, it would seem to me that the valve trays would always leak. Why not?"

"In theory, what keeps the valve trays from leaking is the pressure drop of the vapor flowing underneath the cap."

"Why, Norm, is there a pressure drop of the vapor flow through the cap?" Liz asked.

"Because the vapors accelerate through small holes in the tray deck. Whenever a vapor accelerates, that requires energy. And that energy comes from the pressure of the flowing vapor. Hence the pressure drop. That pressure drop of the vapor is very small. Just a fraction of a single psi. Just high enough to overcome the weight of the liquid on the tray deck, due to the weir height, plus the crest height of the liquid that is overflowing the weir, plus the weight of the cap," I explained to Liz.

"And," Liz said, "if the vapor flow pressure drop is then not big enough, then the liquid will leak through the lower points of the trays. And vapor-liquid mixing efficiency will be ruined and destroy the whole purpose of the tray," she concluded.

Types of Distillation Trays

There are three general types of trays:

1. Bubble caps
2. Sieves
3. Valves

Bubble caps (see Figure 10.1) are currently used only very rarely. They consist of a small chimney with a cap (or a cup) about twice the width of the chimney and perhaps 1" to 2" taller than the weir. Because the chimney is quite a bit higher than the outlet weir, the tray cannot leak any liquid through the cap.

Bubble cap trays only have 90% or less capacity than valve or sieve trays. They are much harder to clean during a plant turnaround than valve or sieve trays. Also, they are much more expensive. Hence, they have been rarely used since the 1960s.

Sieve trays are simply a tray deck with ½" or ¾" holes. They are very simple to clean and less expensive than other trays. Unfortunately, sieve trays have a great tendency to leak if installed out-of-level. And thus will develop quite low tray fractionation efficiency. They also are rarely used in refineries and petrochemical plants, because of these common problems.

Valve Tray Operation

The vast majority of distillation tower trays are valve tray decks. There are two types of valve trays:

1. Flutter-type valve caps
2. Fixed valve caps

The flutter-type valve caps (Figure 10.2) have a rounded or rectangular opening. A tiny cage, or typically one inch (1") long legs, keeps the caps in place. The rising vapor flow causes the caps to be pushed up. If vapor flow is very small, the caps will drop back, to keep the tray deck from leaking.

The flutter-type caps are an engineering error. They are rarely used on new installations, mainly because the cap gets stuck in a wide-open position, or stuck in a closed position, due to fouling. I hate such caps and will no longer use them in my tray designs.

Fixed-Type Valve Caps

Most new distillation towers have fixed valve trays. These valves are formed by cutting two parallel slots on the tray deck, about 1" apart, and typically 2" to 4" long. The slot is then "lifted up" by perhaps 6 mm to 16 mm, depending on the service (see Figure 10.3).

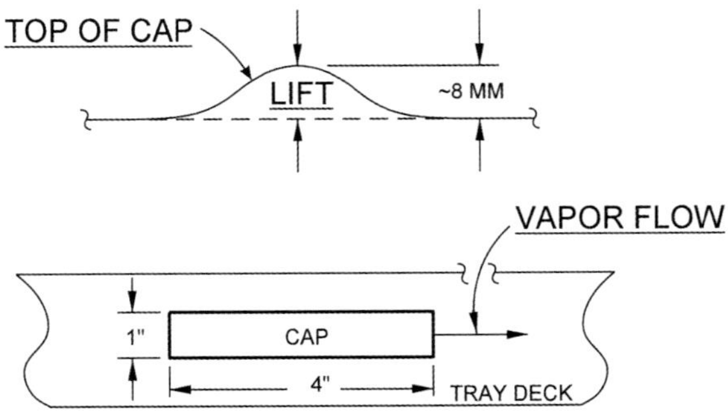

Figure 10.3. Fixed valve cap

The smaller the lift, the less the tendency for tray deck leakage, but the greater the tendency to foul and flood. For clean services, I use a lift of 6 mm. For dirty services, I use a lift of 16 mm. Small lifts should improve tray vapor-liquid contacting efficiency, by reducing the tray deck liquid leakage.

Lift Specification

I always specify the tray deck lift myself. If sections of a distillation tower are known to foul, I'll specify a valve lift of 16 to 18 mm (i.e., perhaps ¾" max lift)

If sections of a tower are certain to be clean, I'll specify a valve lift of 6 to 8 mm (¼")

In practice, I have not observed any large difference, at least in crude distillation services, of tray fractionation efficiency due to different valve cap lifts.

Trayed vs. Packed Distillation Towers

The ability to fractionate efficiently in packed towers is less certain than in trayed towers, due to distributing problems – not so much of the up-flowing vapors but of the descending reflux flow. It is usually the multi-layered liquid reflux distributors that have either not been installed as per design, or not been installed level, that are the problem.

Using a "Radiation Scan" that measures densities at different angles in a packed bed clearly defines the often very large degree of reflux maldistribution on the top few inches of the packed bed of rings, or structured packing.

My main problem with packing in distillation services is that I cannot inspect the internal liquid reflux distributor AFTER it is installed. I have to trust people I don't know to install the complex, multi-layered liquid reflux distributor correctly. This violates a principle I learned from my mother: *DON'T TRUST STRANGERS!*

In my long (60+ years) career of retrofitting refinery crude distillation towers, I have likely replaced trays in the upper sections (i.e., top 4 to 6 trays) in 20 to 30 fractionators that had movable caps, with 1" legs, with fixed valve cap assemblies, with a valve cap lift of 12 mm to 14 mm, so as to avoid tray deck FLOODING, due to the caps sticking to the tray decks, or due to sulfide corrosion products on the top few trays.

Figure 10.4 illustrates the main problem of the loss of tray fractionation efficiency, which is tray deck leakage, due to the tray deck being out-of-level by more than ⅝" to ¾." I use a laser level light, during a distillation turnaround, to measure tray decks for out-of-levelness, in the larger-diameter distillation towers.

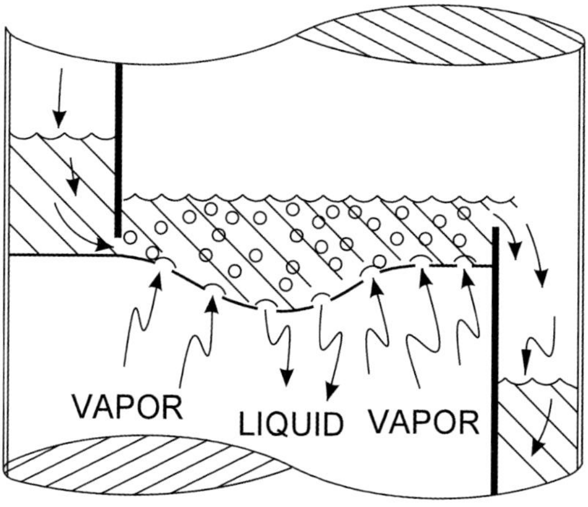

Figure 10.4. Loss of tray efficiency due to out-of-level tray decks leaking

Summary

The selection of the valve cap type is a decision that the process engineer has to make during the design stage of the project. Since the 1960s, bubble cap trays that cannot leak are only used in fewer than 1% of distillation towers.

Valve trays, with movable caps, were typically used from the 1960s through the 2000s. But these caps stuck to the tray deck and/or were blown off.

Currently, fixed valve cap assemblies are used with a lift (i.e., the space underneath the cap) of 6 mm (clean service) or 14 to 16 mm (dirtier services).

Fixed valve cap trays are more robust than those with movable valve caps. A 10-gauge tray deck thickness is my standard, which is much thicker than the 14-gauge industry standard. For corrosive services, I use 316 (or 317) stainless steel, in refinery crude unit services.

Evolution of Trays

Figure 10.1. shows an ancient design for bubble cap trays, first used in the 1870s "Patent Still" to produce brandy from wine. It was still common in the 1960s, when I started my career in Indiana.

The movable valve caps (shown in Figure 10.2) would often become jammed open, or break off, or stick to the tray deck. They have largely been replaced by the fixed valve cap tray in newer towers, as shown in Figure 10.3. However, almost all older towers in the refineries I work in, in Louisiana and Texas, have this sort of inferior movable valve cap tray deck.

11

PICKET WEIRS AND LOW WEIR LOADING

Perhaps the most common design error I have seen in tray design relates to the excessively low weir loadings. Weir loading is the liquid flow over the top edge of the weir, expressed in:

$$(\text{GPM}) \div [(\text{weir length}) \cdot (\text{inches})]$$

To calculate the height of liquid overflowing the top edge of the weir:

$$(0.40) \cdot (\text{GPM/inch})^{0.67} = \text{inches of liquid}$$

That is the "Francis Weir" formula.

If the weir is out-of-level by a few percent, and the tower diameter is small (perhaps 3 or 4 feet), the weir being out-of-level by a few degrees is of no great consequence. But I have designed an 18-ft. I.D. hydrocarbon (delayed coker) fractionation for a refinery in California. There, the center downcomer for the two-pass trays was 17 feet long! A slight out-of-level for such a long weir would result in a very substantial out-of-levelness along the length of the weir, compared to the weir height of 2½".

Pumparound and Fractionation Trays

For the coker fractionator I described above (Figure 11.1), about 60% of the trays were used to fractionate between:

- Naphtha
- Distillate
- Gas oil (light and heavy)

Another 25% to 30% of the trays were in pumparound (i.e., heat removal) services.

But the three trays above the feed inlet (as shown in Figure 11.1) were devoted to removing droplets of entrained liquid in the vapor rising from the vapor space from the flash zone. These droplets of liquid would have a high content of entrained metals, found in the crude charge asphaltene content:

- Nickel
- Vanadium

These metals will poison the downstream hydrocracking catalyst.

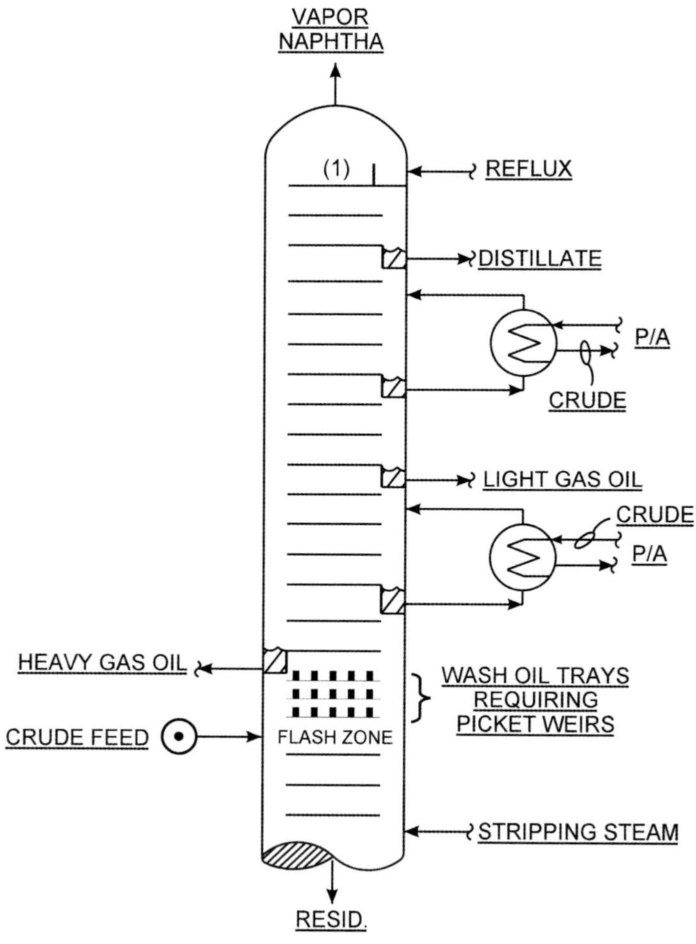

Figure 11.1. Picket weirs

The nickel and vanadium are imbedded in the structure of giant hydrocarbon molecules called "Asphaltenes." These giant molecules are not liquids, but solids. They have no vapor pressure and do not vaporize. The only way the metallic ions of nickel and vanadium can contaminate the "Heavy Gas Oil Product" (Figure 11.1) is by the entrainment of droplets of liquid asphaltenes. The function of the three trays shown with the picket weirs is to wash out these droplets of asphaltenes from the rising flash zone vapors.

Nickel and vanadium contamination of gas oils to be hydrotreated for desulfurization and/or hydrocracking is one of the major process problems for any refinery. Therefore, these metals need to be removed by the three wash oil trays (Figure 11.1), by contact with the down-flowing reflux. But if this wash oil is not properly distributed, the entrained metal (Ni + Va) components can rise into the heavy gas oil draw. This shortens catalyst life and activity.

Using excessive internal reflux, the hydrotreater reactor's would reduce the gas oil recovery. Therefore, to allow a moderate reflux flow, but without causing the slightly higher side of the weir to run dry, picket weirs are used (see Figure 11.2). As discussed below, it's best not to assume that the tray vendors will correctly design the required picket weirs.

Picket Weir Description

Figure 11.2. shows a normal picket weir configuration. The height of the pickets should be two or three times the height of the other weirs. If the weir height is a typical 2½", then the pickets should be about 8" high.

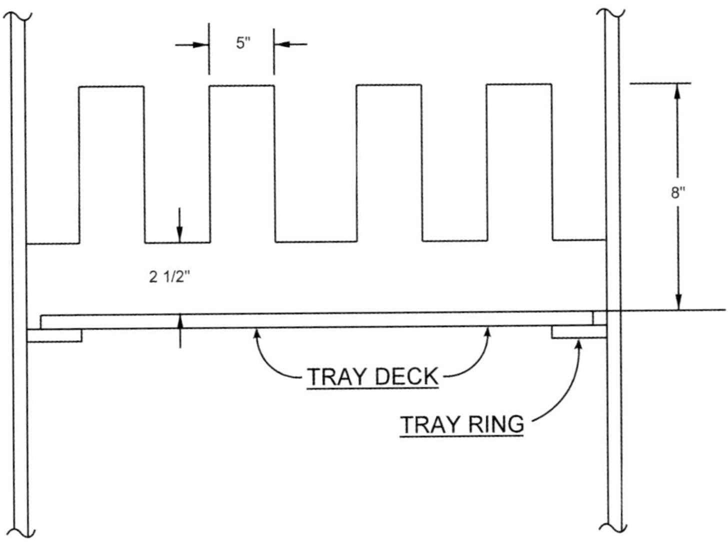

Figure 11.2. Picket weir configuration

I typically make the pickets 4" to 6" wide. The most important dimension is the space between the pickets, which is determined from the Francis Weir formula above, to generate a crest height of roughly 1½ inches – or a weir loading of about 4 to 5 GPM per inch of weir length, between the pickets, assuming that the tray decks and the weirs are LEVEL.

Result of Picket Weirs

As someone who has designed, retrofitted, and reviewed the design of over a hundred coker fractionators, visbreaker fractionators, and crude distillation towers, I am always amazed as to how often picket weirs are omitted by the major suppliers of fractionator tower trays. Their omission of picket weirs on the three, four, five, or six trays above the feed inlet nozzle is typically just an oversight.

When I retrofit such installations, the original tray vendors never object to the addition of picket weirs, which are needed to maintain an even flow of liquid across the tray deck and to remove entrained metals from the up-flowing hydrocarbon vapors.

Reviewing Tray Design

After 60 years as a practicing Refinery Process Design Engineer, I am still amazed as to how often well-established principles, such as the use of picket weirs above the flash zone in crude towers, are omitted. Why I have to remind the tray vendors of their need in refinery crude towers is still a mystery to me. The tray vendors never dispute that they are needed!

The picket weirs do not add any cost to the trays and do not require any additional installation time or installation cost. They certainly do not compromise the trays' mechanical integrity. I suspect their omission is simply a result of using computer technology to design the trays, as it does not recognize the need to maintain a minimum weir crest height for liquid cross-flow on tray decks, with very low liquid flow rates.

Summary

The need to identify which trays should have picket-type weirs in a crude distillation tower is an illustration that we cannot simply accept the tray design, without review, as provided by the tray vendors. We should review the proposed tray design for new distillation towers, to assure ourselves that the new trays are suitable for the intended low weir liquid loading, and other process criteria.

I have found the omission of picket weirs in crude distillation towers to be a very common error in several new crude units, which compromises fractionation efficiency between the diesel oil and the gas oil products. This represents a major economic loss for my clients, in that it downgrades the valuable diesel oil to a lower-value gas oil product.

12

TRAY PRESSURE DROP SURVEY & ANALYSIS

Let us assume that I have a distillation tower with 20 fractionation trays (Figure 12.1) on which I wish to conduct a pressure drop survey to troubleshoot its poor operation. I will measure the pressure drop across these trays by using a single pressure gauge at two points:

- Point #1 – On top of the level glass, at the bottom of the tower.
- Point #2 – At the vapor vent, on the top head of the tower head itself (not on the tower top vapor line). A threaded flange will be needed, to place the pressure gauge on the top head atmospheric vent valve.

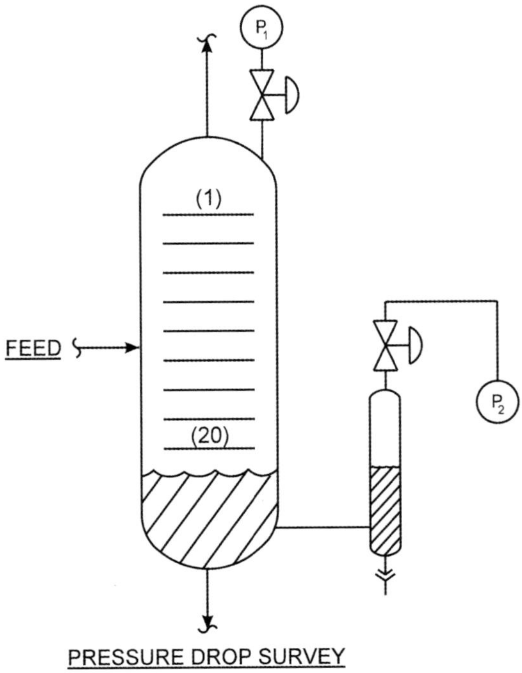

Figure 12.1. Distillation tower with 20 fractionation trays

Be careful to place the pressure gauge in a vertical position at both points. Also, check that the tower pressure has not changed while you are making these measurements. The measured pressure drop, we will assume, across the tower's 20 trays is 3 psi. Let's convert this pressure drop to inches of liquid, as follows:

- Step 1 – Convert the 3 psi into inches of water. There are 28" H_2O for each psi:

$$28 \times 3 = 84\text{" } H_2O$$

- Step 2 – The specific gravity (S.G.) of the hydrocarbon used at the average tower temperature is 0.70 S.G. Therefore, the pressure drop, per tray, in inches of hydrocarbon liquid is then:

$$84\text{"} \div 0.70 \div 20 \text{ trays} = 6\text{" liquid pressure drop per tray}$$

Next, calculate the pressure drop per tray, as a percentage of the tray spacing (the height between the tray decks). Which, for this tower, is 24":

$$6\text{"} \div 24\text{"} = 25\% = 0.25 \qquad \text{Equation 12.1}$$

What does that tell us about the tower's tray performance? Is the 25% too high? Too low? Or just about what we would like for good tray fractionation "efficiency"?

Based on many years of troubleshooting experience, I have developed the following correlations for:

- Valve trays or sieve trays (not suitable for bubble caps).
- In hydrocarbon fractionation services (naphtha, kerosene, gas oil).
- For trays on 18" to 30" tray spacing between the tray decks.
- For a stable operating pressure (i.e., tower top pressure should be steady). Feed and pumparound rates not changing.

The following correlation has no theoretical basis, but is based on empirical observations made by myself, and is summarized below (see equation 12.1, above):

Table 12.1. – Pressure Drop per Tray – Correlation for Valve Trays

1. Normal for good tray fractionation efficiency = 0.18 – 0.28 (i.e., about 0.25)
2. Loss of tray efficiency due to entrainment (i.e., flooding) = 0.30 – 0.45
3. Fully developed flooding = 0.50 plus, or higher
4. Loss of tray efficiency due to tray decks leaking = 0.10 – 0.15
5. Tower completely full = 1.00
6. Tower completely dry = 0.00

Origin of Table 12.1

The correlation parameters shown in Table 12.1 (above) have no theoretical basis. They are purely empirically derived from my field observations. For example, a plant operator tells me, "Mr. Norm! I've adjusted my distillation tower top reflux rate and bottoms reboiler duty to maximize the separation efficiency in my propane-butane splitter."

I will then measure the percentage as I've shown in equation 12.1 above, which should yield a result of between 18% to 28%, for ordinary valve cap type trays. Why? Because the operator has optimized reflux and reboiler rates to maximize the fractionation efficiency in his distillation tower, and I only then observe the tray pressure drop that corresponds to his field adjustments.

On the other hand, if I measure a pressure drop through a 40-tray fractionator of essentially zero and next open the tower top vent, and hydrocarbon liquid squirts out, the problem is "Top Tray Flooding" due to salt and/or corrosion product accumulation on the top few trays of the tower. This is potentially dangerous! Do not engage in this type of reckless behavior just to solve a process problem.

Pressure Drop Distribution

The above correlation has a potential fundamental problem. It assumes that the pressure drop of the flowing vapor is equally distributed across the trays. Sometimes, however, it is only the top few trays that are flooded, due to fouling with salts and/or corrosion products. This is a common problem in hydrocarbon distillation towers, where moisture is formed due to the condensation of the water vapor in the vapors. The water often condenses on the top few trays, and this causes aqueous phase corrosion and fouling. The resulting salts and corrosion products cause a localized high DP, but only on the top few trays. The resulting tray deck flooding progresses up the tower, but DOES NOT progress down the tower.

To mitigate this common problem, I always try to obtain DP measurements over as many intermediate sections of the trays as the available pressure taps permit.

The Safe Way to Make a Pressure Reading with a Pressure Gauge

- Step 1 – Remove the small screw at the bottom of the gauge's threaded nozzle.
- Step 2 – Close the ¾" valve on the pressure tap tightly, with a wrench.
- Step 3 – Remove the ¾" threaded plug.
- Step 4 – Install any fittings required to install gauge in an upright position.
- Step 5 – Do not tighten the gauge by twisting by hand. Use a wrench to tighten.
- Step 6 – Open valve 100%. Note pressure.

- Step 7 – Close valve 100%, and remove gauge.
- Step 8 – If gauge does not return to zero, pressure tap cannot be used without clearing the tap with a rod out tool. Do not use a bent welding rod.
- Step 9 – Replace gauge and open valve 100%. Note pressure. If readings at Step 6 and Step 9 are not the same, the pressure tap obtained from the plant's safety department. Cannot be used without clearing tap with a rod out tool.

Summary

The rod out tool is a standard piece of safety equipment, usually available in most process plants, used to clear small (½" or ¾" or 1") fouled bleeders. Clearing a plugged bleeder with a bent welding rod—a common practice, which I have done often myself—is a very dangerous practice.

I have often been asked by attendees in my seminar, after 60 years of refinery distillation tower troubleshooting, why I still perform fractionator pressure surveys myself? My response is that for me to correctly and confidently identify the cause of poor tower fractionation efficiency, I have to be able to have confidence that the tower pressure drop survey has been done carefully. And directly observing or executing the survey personally is the only method I have to guarantee the result and the accuracy of such pressure drop surveys.

Safety Note

One of the leading causes of fires and deaths in petrochemical and refinery services is due to engineers, like us, running pressure drop surveys, but without taking the proper safety precautions.

1. Certainly, tighten the ¾" or 1" valve handle with a wrench, prior to pulling the plug in the valve's screwed fitting.
2. Then, loosen the plug with your wrench, until you can wiggle the plug with your fingers.
3. If the plug leaks the very slightest amount, STOP! You need to alert the unit supervisor of a potentially hazardous situation.
4. Next, pull the plug, and screw in your pressure gauge tightly.

Open the valve. If the needle on your pressure gauge does not increase within a few seconds to its final pressure, do not clear the valve with a bent welding rod. It requires the use of a "Rod Out Tool." Safety first!

13

MECHANICAL ASPECTS OF TRAY EFFICIENCY

This chapter deals with how the installation and the mechanical design of distillation trays affect the tray's fractionation efficiency. Regardless of how well the tray is designed, if it is not installed carefully, or lacks sufficient structural strength, its fractionation efficiency will be compromised.

The problems described below are not critical for a small-diameter tower of less than 4 ft. But I have designed 40 ft. diameter vacuum towers, and the problems described are more critical in achieving good tray fractionation efficiency in larger-diameter towers.

Tray Deck Design

The majority of distillation towers rely on perforated tray decks to mix the up-flowing vapor with the down-flowing liquid reflux. Older distillation towers were equipped with bubble cap trays. Actually, some quite new towers are so equipped. Bubble cap trays are superior to the perforated trays used in 99% of new distillation towers. But 99%+ of the towers I encounter have:

- Valve trays – movable caps, or
- Sieve trays – trays with holes, or
- Grid trays – fixed valve-type caps.

I will skip discussing inspecting bubble cap and the similar tunnel cap tray decks, and only discuss perforated tray decks (i.e., valve, grid, and sieve trays).

I'll discuss inspection of fractionation towers using beds of structured packing in a subsequent chapter. This is a very complex subject that is best avoided by not using structured packing in fractionation services (see Chapter 17).

Distillation tray components are shown in Figure 13.1. The tray deck is covered with valve caps, slats, or holes. The vapor rises up through the caps or holes and mixes with the liquid on the tray deck.

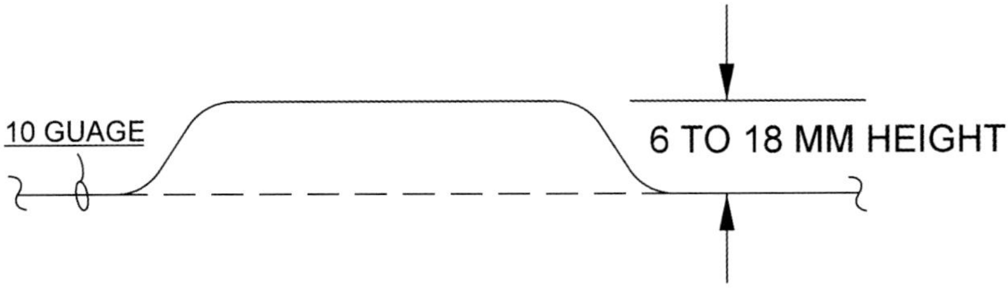

Figure 13.1. Distillation tray components

The height of the liquid on the tray deck is controlled by the weir height (Figure 13.2). The pressure drop of the vapor flowing through the holes, or valves, on the tray deck is, by design, approximately equal to the weight of the liquid on the tray deck. Otherwise, the tray deck would leak.

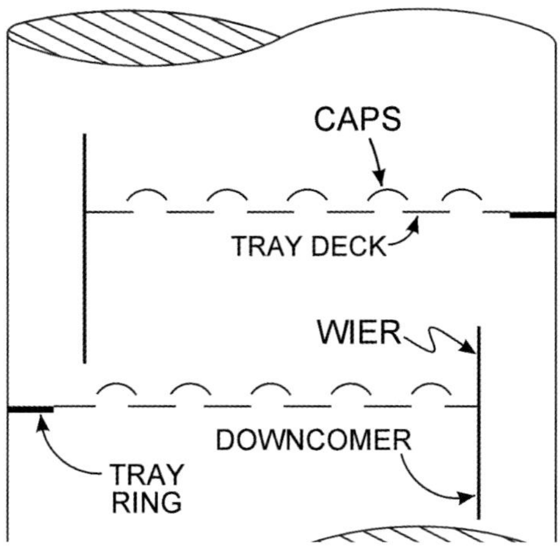

Figure 13.2. Slotted tray decks

The weight of the liquid on the tray deck is called “Delta P Wet.” The pressure drop of the vapor flowing through the tray deck is called “Delta P Dry.” By design, Delta P Dry approximately equals Delta P Wet.

Variations of this ratio of one-to-one (Delta P dry vs Delta P wet) by 30% to 40% are acceptable. That is, the vapor and liquid are still able to mix pretty well. But if the tray deck is out-of-level by more than one inch, on a tray with a weir height of 2” to 2”, then:

- The liquid will leak through the low points of the tray deck.

and

- The vapor will flow through the high points of the tray deck.

The net result will be that the vapor will not come into intimate contact with the liquid. Vapor-liquid mixing efficiency will be poor. The tray's fractionation efficiency will be degraded, and the separation between components will be degraded.

According to a major tray vendor, the maximum allowable tray deck out-of-levelness is 5/8" (0.625"). In a small-diameter tower, such as 36" I.D. depropanizer, this 5/8" out-of-levelness is not difficult to achieve. But, in the 18-ft. I.D. delayed coker fractionator I designed for a refinery in California, a 5/8" out-of-levelness specification was really difficult to achieve.

Inspecting for Tray Deck Levelness

I have used a laser level light to check for tray deck out-of-levelness. To some extent, the tray installation technicians can correct the tray out-of-levelness for newly installed trays. In towers that have been in service for years, the chances of getting tray deck levelness upgraded are rather small. On the other hand, weirs that are out-of-level can be adjusted to a good degree of levelness, with a reasonable effort.

The time to improve tray deck levelness potential is when the trays are first ordered, as discussed below.

Improving Tray Deck Levelness

The single best way to ensure tray fractionation efficiency by enhanced tray deck levelness is to purchase trays with an "Integral Truss," as shown in Figure 13.3. This type of tray assembly greatly increases tray deck levelness (also called "Back-to-Back" trays).

The other way to improve tray deck levelness it to make the trays out of thicker metal. In the 1960s, the industry standard for tray deck thickness was 16-gauge. The larger the gauge rating, the thinner the tray deck.

After long experience of tray failures, tray vendors increased the tray deck standard thickness to 14 gauge. This is roughly 15% thicker than 16-gauge. My standard to promote tray deck levelness is to specify that 10-gauge thick trays are required. This is approximately 50% thicker than the 16-gauge tray decks used in the 1960s.

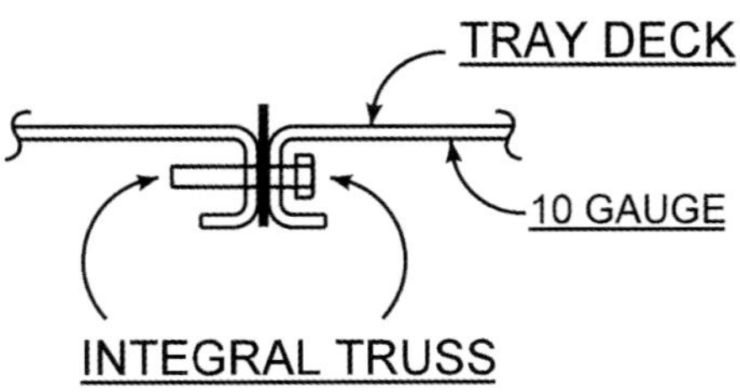

Figure 13.3. Back-to-back trays

Fixed Valve Cap Tray Decks

Trays with movable valve caps require that a 1½” hole be cut through the tray deck. This weakens the tray deck.

Sieve trays also require that a ½” or a ¾” hole be drilled through the tray deck. This also weakens the tray deck to some extent.

The newer types of trays have openings for the vapor flow by:

1. Cutting 2” parallel slits in the tray deck, several inches apart.
2. Pushing up the area between the dual slits to a height depending on the design vapor flow rates (6 mm to 16 mm).

The height of the slot (see Figure 13.2) depends on the vapor flow rate and the fouling tendency of the service. A minimal height for a very clean service is 6 mm (i.e., ¼”). A maximum height for a slot, for very dirty services, would be 16 mm (i.e., 2/3”).

These narrow slots have a much smaller tendency to leak than do the 1½” round holes used for valve trays. The reduction in tray deck leakage, and improved tray integrity, greatly improves tray fractionation efficiency, based on my long experience.

Replacing Tray Deck Manways

One of the more common problems I have encountered, leading to a loss of distillation tray efficiency, is that during a turnaround, the tray deck manways are not all replaced.

What happens is that the tower is approved for closure after the final inspection. The contractor personnel then are supposed to reinstall the tray deck manways. But if no refinery personnel are checking this manway reinstallation, then often only the tray deck manways above and below the tower external manways are replaced. Once these tray deck manways are bolted down, there is NO ACCCESS to inspect the other tray deck manways for closure.

To avoid this very common problem, the unit engineer should inspect each tray deck manway after it is installed. Once a tray manway is secured, there is no access to the trays below for the final inspection.

Field Testing for Tray Decks Out-of-Levelness

The following procedure is what I’ve done several times in fractionation units, mainly in large-diameter, delayed coker fractionators (I have designed five of these).

- Step 1 – Maintain a constant reflux rate and a constant reboiler duty throughout this test.

- Step 2 – Reduce the tower top pressure by perhaps 1% to 2% of the tower's normal operating pressure. This assumes that you have excess condensing capacity. You may have to do this test at night when air or cooling water temperatures are lower.
- Step 3 – Observe what happens after the tower lines-out (20 to 30 minutes) to the Tower's D.T. temperature (D.T. = Tower Bottom – Tower Top).
- Step 4 – If this differential temperature increases by a few degrees, then repeat Step 2 above.
- Step 5 – Continue with this test until you run out of condenser capacity or the tower differential temperature (between bottom and top) begins to diminish.

Conclusions for the Above Test

The main reason for a larger differential temperature (which does indicate improved fractionation efficiency) between the tower top and bottoms is improved tray vapor-liquid contacting efficiency.

The main reason why a lower tower pressure (as per above test) would improve fractionation efficiency is reduced tray deck leakage at the low points of out-of-level tray fractionation decks (Figure 13.4).

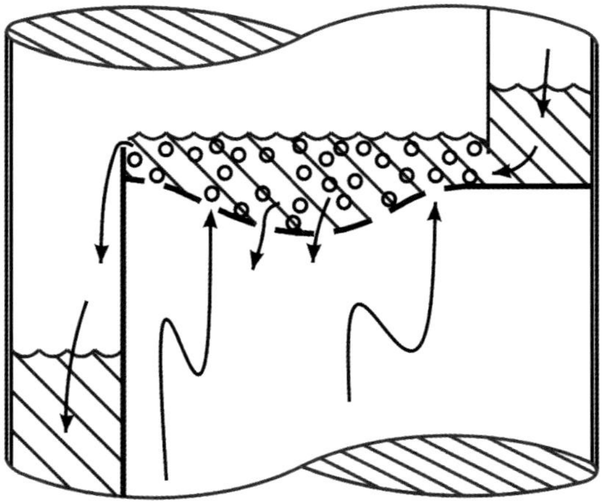

Figure 13.4. Out-of-level tray fractionation deck

We do wish to operate our fractionators at the optimum pressure, which gives the desired degree of fractionation at the minimum reflux ratio, to minimize reboiler duty. The observation that lowering tower pressure also increases vapor velocity, vapor pressure drop through the tray deck, and improves tray fractionation efficiency is a positive indication of:

- Tray deck leakage
- Tray decks out-of-level
- Tray decks that need to be re-leveled or replaced

Downcomer Bracing Brackets

In large-diameter towers, the downcomers will be very wide. This results in a side downcomer that is also very wide. As the downcomer may be 36" long and constructed out of thin 16-gauge metal, it will be quite flexible and thus subject to distortion. The pressure drop of the vapor, pushing up through the tray above, may also push the bottom edge of this long, flexible downcomer towards the vessel wall.

A loss of downcomer area may result at the bottom edge of the downcomer. This is a common cause of downcomer backup and of flooding of the tray above. To avoid this problem, we use small clips fixed to the edge of the downcomer and to the tray deck below, spaced about three feet apart (Figure 13.5). When I inspect trays in large-diameter distillation towers, I always check that these clips are tight and in place.

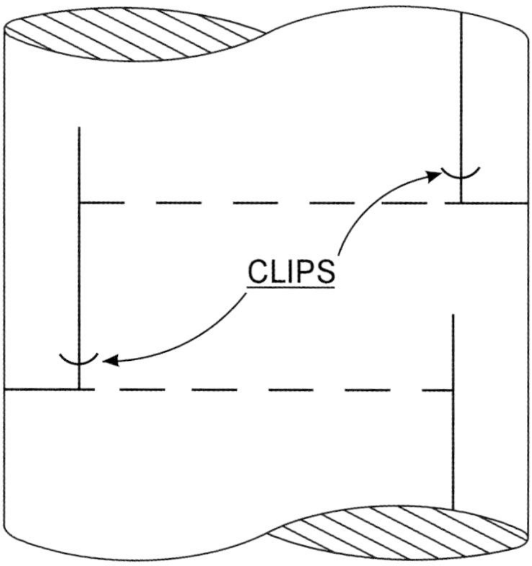

Figure 13.5. Downcomer clips

Tray Levelness Upgrade – Mechanical Specification Details Summary

In order to improve installed levelness of trays, I specify:

1. Back-to-back trays, with an "Integral Truss"
2. 10-gauge tray deck thickness (not 14-gauge)
3. Fixed valve cap assembly
4. Shear clips, connected to the integral truss (slide underneath the tray rings)

Tray Metallurgy

If virgin crude oil is to be distilled, then naphthenic acids are almost certainly going to be present. To avoid corrosion, I would then specify 316 (2½% Moly) stainless steel. Or, better yet, 317 (3½% Moly) stainless steel.

Downstream of the crude unit, after products have been processed in:

- Fluid catalytic cracker
- Delayed coker
- Visbreaker
- Hydrotreater

I would use 410 chrome stainless (which is far less costly than 316 or 317 stainless steel), as all the naphthenic acid would have been destroyed.

Corrosion of trays due to HCl is a more serious problem, which may require use of Hastelloy metallurgy for the trays.

Summary

Regardless of how well the process configuration of a distillation tray is designed, if it is not mechanically robust, and not installed correctly, it will not fractionate properly. In my long experience, the most frequent and the most detrimental of tray installation errors is that the tray deck manways are not reinstalled after a unit turnaround. Tray inspection, to check for levelness and mechanical assembly, according to the tray's design, is best carried out personally by the unit's process engineer and lead operators, after the unit turnaround has been completed.

Bubble Cap Trays

The bubble cap tray was used in the early distillation towers in the 1870s, to produce brandy from wine. The bubble caps used in this early distillation tower had a slotted hat that covered a chimney. The height of this chimney was 1″ to 2″ higher than the height of the outlet weir. Hence, the tray could not leak.

The tray had extremely rugged construction. The few times I have been in a tower to inspect very old bubble cap trays, they were still in perfect mechanical condition.

14

DISTILLATION TRAY – FLOW PATH LENGTH PROBLEMS

My first project as a Process Design Engineer was to increase the propylene recovery from a Fluid Catalytic Cracker gas plant. The propylene recovery was only 60%, and I was assigned to increase it to 80%.

Recovery of an LPG component, such as propylene, from a vapor stream, is accomplished by circulating a heavier lean oil liquid hydrocarbon stream, consisting of a heavier naphtha, to absorb the propylene from the flowing gas stream in an absorption tower. This will increase "A," the absorption factor:

$$A = L/VK$$

Where:

V = Moles of vapor

L = Moles of lean oil

K = Vapor pressure of propylene (psia) ÷ Tower operating pressure (psia)

Weir Loading Limit

The problem with increasing the lean oil flow was the liquid loading on the overflow weir (Figure 14.1) exceeded the "140 GPM per ft." of weir length allowable weir loading in the published industry standards for tray design.

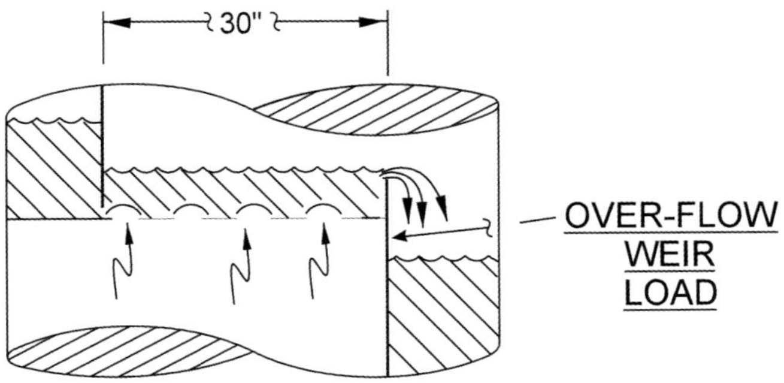

Figure 14.1. Excessive weir load

To overcome this limit, I decided to convert the propylene absorber trays from the single flow path shown in Figure 14.2 to the two-flow path shown in Figure 14.1. This would reduce the weir loading from the excessive 140 GPM/ft. to a more normal 80 GPM/ft.

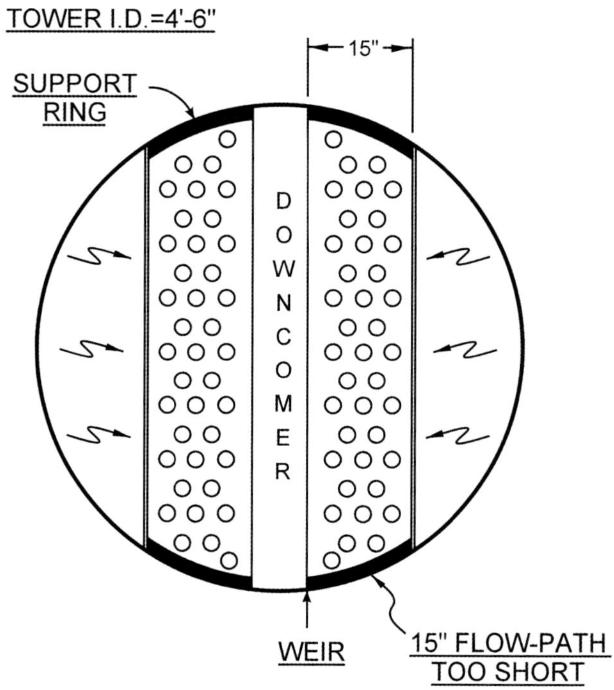

Figure 14.2. Two flow path

Result of Revamp

About a year passed. My project was approved by management. The necessary, larger, lean oil circulation pump and piping were designed, purchased, and installed. The new, two-pass trays I had designed were also installed in the propylene absorber tower.

When the propylene absorber started back up, Jerry, the Unit Process Engineer, phoned me, "Hi Norm! I hate to tell you this, but our former low 70% propylene recovery is now only 60%!"

A totally catastrophic result for my very first process design project. Not a good way to begin my engineering career.

What Had Gone Wrong?

I had not understood the factors that affected tray vapor-liquid contacting efficiency. One of these factors is the tray's "Flow Path Length." This is the distance between the downcomer from the tray above (Figure 14.1) to the overflow weir. This was 30" before my tray modifications.

But, as shown in Figure 14.2, with the two-pass trays, this flow path length had been reduced to only 15". Because of liquid by passing around the tray support ring (shown in Figure 14.2), a reduction in the tray's flow path length to less than two feet (24") had resulted in poor vapor-liquid contacting efficiency, and thus reduced propylene absorption.

When I converted from the single pass trays in Figure 14.1 to the two-pass trays shown in Figure 14.2, I had more than DOUBLED the amount of lean oil that was by passing the vapor flow around the edges of the tower devoted to the tray support rings.

LESSON – Do not design distillation trays to have a distance between the downcomer from the tray above to the tray's outlet weir to be any less than 24", to avoid the loss of tray efficiency. If you cannot meet this criteria, you are better off using a "Packed Tower" for your design for small-diameter distillation towers (as discussed in subsequent chapters in this book).

My Published Paper

The rest of this true story follows:

I had been committed to publish the result of my project by management. So, my paper describing this project was published in *Chemical Engineering Progress*, Volume 64, No. 7, July 1968. The title of the article was "Bottleneck Removal in a Reboiled Absorber." I produced a plot of:

Absorber Vapor Flow Rate vs. Propylene Recovery

The plant data before and after my design was implemented, was provided to me.

The propylene recovery after my retrofit design was POOR. So, I erased the data points that showed the worst propylene recoveries (see Figure 14.3 in my 1968 publication). But other than the Unit Engineer, Jerry, no one ever commented, or even noticed, my technical duplicity.

Minimum Flow Path Length

Always maintain in a trayed distillation tower a minimum flow path length of at least 24". One way I have used to partially offset the loss of tray efficiency for very short flow path lengths is to install small (2" wide, by 3" high) dams around the periphery of the tray. This reduces the tendency of the cross-flow liquid to partly by pass the "Bubble Areas" (that is, the active areas) of the tray decks, and this will improve the tray's vapor-liquid contacting efficiency.

A Lesson Learned

I do have a real talent NOT to repeat my design errors in applying distillation technology. The other lesson to be learned from this incident is not to rely on the distillation tray vendors to avoid making tray design errors. Their business is to sell trays. It is not to avoid making tray process design errors. As I learned very early in my career, there are only two people you can trust: yourself and your mother. And, unless your Mom is a Process Design Engineer, you will have to learn to rely on yourself, as I have.

Improving Efficiency in Small-Diameter Towers

For towers of less than 42" I.D., one way to improve fractionation efficiency is to drill holes through the tray support rings. This ring is typically 2" to 2½" wide. A significant percentage of the down-flowing liquid will then by pass the up-flowing vapors around the periphery of the tray deck. Drilling ½" holes, on 2" to 3" spacing, will then significantly enhance vapor-liquid contacting efficiency, by diminishing the "Edge Effect" (i.e., tray hole area by passing around the tray support ring).

This can only be done after each tray is installed. This concept does apply to two-pass tray decks, with short (less than 18") flow path lengths, in larger-diameter (60") towers.

The placement of 2" high by 2" wide dams, along the periphery of a tray, with short flow path lengths is also an acceptable practice. But this may slightly reduce the tray's capacity, as compared to the option of drilling holes in the tray ring, which will increase a tray's capacity, in smaller-diameter distillation towers.

For very small-diameter towers, the tray vendors sell pre-assembled inserts that can slide into a small diameter tower without any installation problems and without any tray rings.

Summary

Any distillation tray with a flow path length of less than about two ft. is likely to suffer from low distillation efficiency. Blocking off the tray area around the edge of the tray, on top of the tray ring, will mitigate this loss of fractionation efficiency.

15

ON-STREAM WATER WASHING OF DISTILLATION TOWERS

The most common cause of flooding on the larger refinery trayed distillation towers, is salt sublimation on the upper few trays of:

- Crude Units
- Fluid Catalytic Cracking Units (FCU)
- Delayed Coker Fractionators
- Visbreaker Fractionators
- Vacuum Towers

The salts sublime out on the tray decks. The term "sublime" means that a vapor precipitates as a solid, but without first becoming a liquid. These salts originate from the sea water salts that are dissolved in crude. Sea water salt consists of approximately:

- 95% NaCl
- 3% $CaCl_2$
- 2% $MgCl_2$

The NaCl does not dissociate at temperatures below 800°F, and is therefore not a problem in crude units. But the $MgCl_2$, and to some lesser extent the $CaCl_2$, will dissociate in the crude units to release hydrochloric acid (i.e., HCl). There is also nitrogen in the hydrocarbon, which can form NH_3. And of course, there is a lot of sulfur in the crude, which will form H_2S. This mixture of:

- HCl
- NH_3
- H_2S

formed in an aqueous environment will foul the tray decks with salts and corrosion products. This will then cause the upper trays to flood, due to excessive pressure drop of the vapor, as it flows through the restricted and fouled tray decks. This is an especially serious problem if valve-type trays, with movable valve caps that have a tendency to stick to the tray deck, are used.

Sublimation of Ammonia

Most of my personal experience with tray flooding, due to salted trays, is related to distillation towers in delayed coker services. There, the common problem is NH_4Cl sublimation on the upper tray decks of the delayed coker, wet gas absorber tower. This tower recovers propane and butane from the delayed coker off-gas stream.

The term "sublimation" is defined as the direct conversion of the salt from the vapor phase into a solid phase (i.e., ammonia chloride crystals or salt). When I worked on the delayed coker in Indiana, in 1969, we used to water wash the delayed coker wet gas absorber perhaps once a week for 20 minutes, or once a month for a few hours, to remove the accumulated sublimed ammonia salt crystals from the tray decks. The objective was to prevent the absorber from flooding. The salts would cause the tray deck valve caps to stick to the tray decks.

Water Washing FCU Fractionators

Much of my personal experience, from a process design perspective, is the design to water wash fluid catalytic cracking unit fractionators. Basically, this consisted of:

1. Taking a slip-stream of water from the main fractionator reflux drum boot.
2. Connecting this water to the fractionator top reflux line.
3. Perhaps halfway between the tower top tray, and the light cycle oil pumparound return, selecting a tray with side downcomers (assuming trays are two-pass).
4. Installing a sump (see Figure 15.1), perhaps half of the tray spacing deep (12" to 15"), and as wide as the center downcomer of the tray above, plus a few inches.
5. Connecting the water drain nozzle to the side of this sump, at its floor (Figure 15.1) and out through the vessel wall; sizing the water draw-off nozzle for approximately 3 ft. per second, to avoid nozzle cavitation, and thus avoiding the need for a vortex breaker, which could foul.

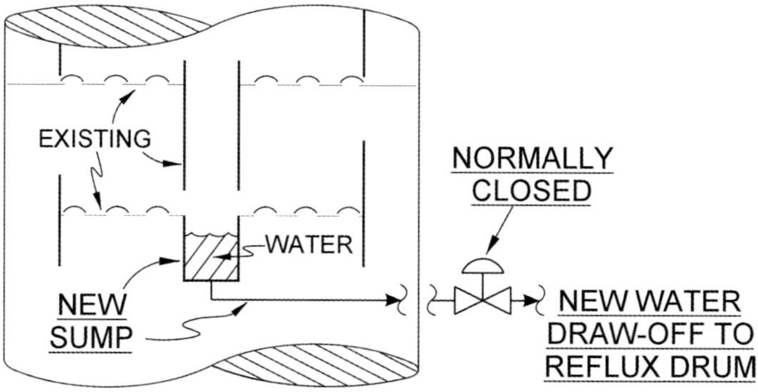

Figure 15.1. Water draw-off tray sump

Water Wash Frequency

The operators on one of the FCUs that I retrofitted with this feature used to wait for their DP across the upper section of the FCU fractionator to start to rise (perhaps every few weeks), and they would then water wash the tray for 20 or 30 minutes, for salt removal.

There was no meter on this wash water. But the wash water flow control valve was operated from the panel. I recall that the wash water was recycled back to the fractionator reflux drum, which was located below the water draw-off nozzle, for disposal.

Desalter Operation

In a refinery, the best way to avoid the problems discussed above is to improve the performance of the crude unit electric grid desalter. To optimize the desalter performance, based on my experience:

1. Keep the desalter temperature somewhere between 270°F and 290°F (never above 300°F, to protect the electric components).
2. The percentage of water based on the crude flow rate should be greater than 6%. Seven percent is better. Four percent is too low. Ten percent was the industry standard in the 1960s, but that was excessive.
3. Inject at least half of the wash water at the inlet of the crude preheat exchanger train, and the rest through the desalter mix valve. Injecting all of the wash water at the exchanger train's first exchanger is acceptable practice.

4. When using sour water stripper bottoms as the desalter wash water, target for an NH_3 content of 20 to 40 ppm. A 100-ppm NH_3 is extremely excessive. Less than 10 to 20 ppm of NH_3 serves little purpose.
5. Keep the desalter pressure at least 20 psi above the calculated vapor pressure of the crude, to avoid vaporization.
6. Control the wash water pH to keep below 9 pH with H_2SO_4, if necessary.

Summary

On-stream, water washing of fluid catalytic cracker and delayed coker distillation tower trays is a standard industry practice, to extend run lengths between turnarounds, when the run lengths have been limited by a progressive loss in distillation tower capacity, due to flooding in the upper 10 to 12 trays in the fractionator, as a consequence of salt formation on the distillation tray decks. The higher the nitrogen content of the crude, the greater will be the tendency to flood the upper trays of the fractionator, due to the salt sublimation. My personal experience for on-stream water washing of trays is extensive and extremely positive, to prevent premature tray deck fouling and flooding. It is best to water wash too frequently than not to water wash the upper distillation tower trays often enough to prevent flooding.

16

UNDERSTANDING VAPOR FLOW IN DISTILLATION TOWERS

What happens to the pounds per hour of vapor as it flows through a distillation tower? Do the pounds of vapor increase, or decrease, from the top to bottom? As everything else in distillation technology, the answer is complex! As you read this chapter, you may wish to look at a "Tray-to-Tray" computer simulation of a multi-component fractionator, such as a naphtha debutanizer. Reading this chapter will help you understand the computer simulation output better.

Starting at the feed tray, the liquid trickles down the tower, into the reboiler. As the liquid flows down to the bottom part of the tower, it absorbs heat from the up-flowing vapors. Some of this heat is the latent heat of condensation of the up-flowing vapors. Therefore, as the liquid flow descends from the feed tray into the reboiler:

1. The flow of the liquid on the trays below the feed tray goes up.
2. And the flow of vapor is reduced as it partially condenses into the descending liquid flow.

Why does the vapor flow decrease as it flows up the tower below the feed tray? Because the latent heat from the vapors leaving the reboiler is partially absorbed by the descending liquid from the feed. And these vapors partially condense into liquid, which then joins up with the descending liquid from the higher feed tray.

On the other hand, above the feed tray, the flow of vapor will increase. But why?

Because the up-flowing vapors cool off as they rise from the feed tray as the vapor flows up the tower to the tower top tray. And what then happens to the heat lost by the cooling vapors as these vapors rise up from the feed tray to the tower top tray?

The answer is that this heat is absorbed by the descending reflux. As the descending reflux flows down the tower, some of this reflux flow evaporates. Therefore, above the feed tray in a distillation tower:

1. The vapor flow increases from the feed tray to the top tray.
2. The liquid flow is reduced from the top tray to the feed tray.

Effect on Tray Flooding

When a distillation tower floods, this detrimental tray flooding is usually caused by the flooding initiated at the:

- Top few trays
- Bottom few trays

Of course, if the flooding is initiated at the tower bottoms trays, the entire tower will flood. But that takes time. Perhaps one or two hours is a rather typical response time for the flooding to develop, in my experience.

Tray Fouling Distribution

Magnifying the problem described above of maximum vapor and liquid loads occurring on the top tray and on the bottom tray in a typical distillation tower, is the distribution of fouling on the tray decks.

Below the feed tray, the tray most likely to experience fouling is the bottom tray of the tower, at least in hydrocarbon distillation services. This is especially true if the feed has been exposed to air, and/or contains di-olefins (i.e., the precursor of gums). Basically, the bottom tray will foul before other trays, because it's always the hottest tray.

Above the feed tray, the tray most likely to experience fouling is the very top tray or top few trays of the tower. This is primarily because of the tendency of water to condense, or salts to sublime, at the lower top tray temperature.

If the bottom tray fouls, given enough time, the entire tower will eventually experience flooding, as the downcomers progressively back up. If, on the other hand, the top few trays flood, due to the sublimation of salts, the flooding will NOT progress down the distillation tower, but liquid reflux will back up into the overhead vapor line, and eventually into the overhead condenser and back into the reflux drum.

Effect on Pressure Drop Survey

Ordinarily, we would identify a loss of fractionation efficiency due to flooding by conducting a field pressure drop survey. If the pressure drop per tray indicated was more than 0.3 to 0.4 psi, we would suspect that our poor fractionation efficiency was a result of tray flooding and downcomer back-up.

But if our problem was top tray flooding, causing a loss in distillation tower fractionation efficiency, then the observed tower pressure drop would be quite low. And it would remain low as we increased the reflux rate, even with the reboiler running on automatic tower bottoms temperature control.

When in doubt, I may barely and carefully crack open the atmospheric vent on the top head of the tower (being 100% careful to avoid any possible source of ignition). If liquid squirts out, then the tower is flooded for sure, even if the measured overall tower pressure drop is extremely low. I always rely on direct and personal field observations as the ultimate test.

Best to obtain management approval before opening the atmospheric vent!

Loss of Downcomer Seal

One of the more common causes of a distillation tower flooding, is the loss of a downcomer seal. How does this happen?

The usual cause is that a tray deck is leaking so badly that the depth of liquid on the trays falls below the bottom edge of the downcomer from the tray above. Vapor will then flow up through the downcomer. The vapor flowing upwards through the downcomer prevents the liquid from draining down through the downcomer. The liquid will then back up the downcomer onto the tray above. This tray then floods! And the flooding progresses up the tower.

To stop this flooding, I will increase the vapor flow in the tower. This will reduce the tray leakage and will restore the liquid level on the tray deck. The best way to avoid this problem is to have an inlet weir of about 3" in height next to the bottom edge of the downcomers, to preserve the downcomer seal from the tray above.

Safety Note

Opening the top vent for a distillation tower in refinery or hydrocarbon services, is a potentially very dangerous action and should not be done without adequate safety precautions, and only with prior approval by plant management. Safety must come first!

Summary

The distribution of the vapor flow through various sections of a distillation tower, is a complex subject, when dealing with multi-component distillation. Contrary to what we might expect, a reduction in the tower top temperature, as a consequence of an increase in the reflux rate that will reduce the tower top temperature, will result in an INCREASE in the vapor flow, by volume and by weight, on the top few trays. This is a frequent cause of flooding in towers in multi-component distillation services.

17

DISTILLATION PROBLEMS USING STRUCTURED PACKING

The following chapters relate to distillation in packed towers using:

1. Structured packing
2. Grid
3. Dumped-type packing (i.e., rings and saddles)

Mainly, these four chapters involve the advantages and the problems of using beds of modern structured packing that are often used in preference to trayed distillation towers to enhance the capacity and the fractionation efficiency of many distillation towers.

A major portion of my consulting business is retrofitting distillation towers that have beds of the structured packing that are malfunctioning due to poor liquid distribution.

A typical bed of structured packing is shown in Figure 17.1. The bed is constructed out of very thin sheets of either:

- 317 stainless steel
- 316 stainless steel
- 410 stainless steel

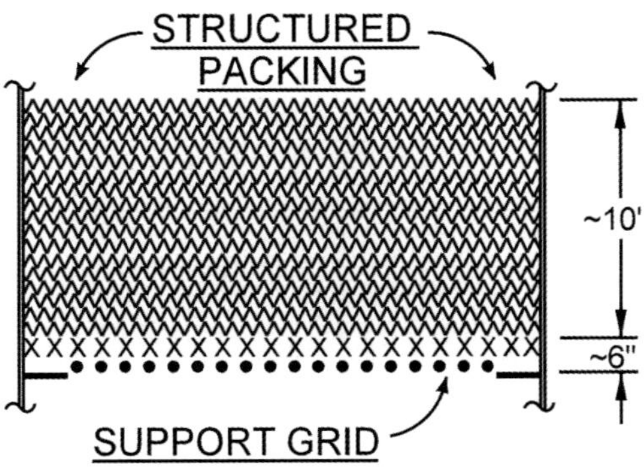

Figure 17.1. Structured packing

The stainless steel (S.S.) sheets are crimped at a spacing of ¾” or 1” and then the crimped, thin sheets are wired or bolted together, to form blocks of structured packing that are typically:

- 12” wide
- 12” high
- 3 ft. to 10 ft. in length (depending on the diameter of the tower)

The beds of packing are sealed against the vessel wall, with very thin rings of bent stainless steel, about 3” in length.

The Advantages of Structured Packing

Structured packing was initially introduced in the refining industry in 1965, as an alternative to trays. The purpose of the packing above the feed nozzle is to de-entrain the up-flowing vapors. Structured packing does an excellent job of removing tiny droplets of liquid from the up-flowing vapors, as long as the initial vapor distribution is good.

Liquid Distribution

The purpose of the wash oil, shown in Figure 17.2, is not to fractionate. It’s only intended to wash off the heavy asphaltene molecules that adhere to the surface area of the structured packing, to prevent them from “coking up,” or fouling, the wash oil grid.

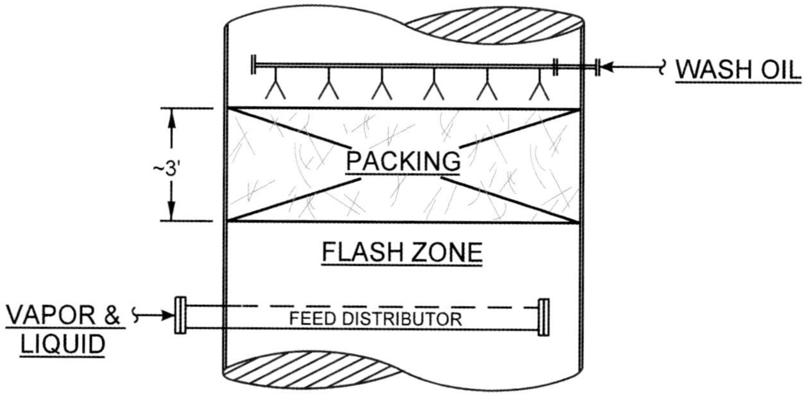

Figure 17.2. Purpose of wash oil

To fulfill this function, spray nozzles are used to distribute the wash oil over the top of the grid. But there is a big problem with the use of spray nozzles that prevents them from being used for liquid distribution on beds of structured packing in fractionation services. There, even distribution of the reflux is mandatory, to obtain a reasonable fractionation efficiency in the bed of packing.

Spray Nozzle Distribution

The main problems with spray nozzles as liquid distributors are that:

1. 80% to 90% of the sprayed liquid is in the outer few inches of the spray cone.
2. Only 10% to 20% of the sprayed liquid is in the center few feet of the spray cone. That is, the interior of the cone is mostly dry.

This is based on tests I have run with ¾", 120°, "Full Cone," spray nozzles, purchased from "Bete Spraying Systems Inc," a leading supplier of spray nozzles used in refinery applications. The 120° refers to the spray angle.

To remove the entrained asphaltenes in the packing in the wash oil grid efficiently, the 120° spray pattern of the wash oil is designed for a 100% overlap with the adjacent spray nozzles. This is an acceptable design for grid wash oil services, but not in hydrocarbon fractionation services.

Use of Spray Nozzles in Fractionation Services

Unfortunately, this distribution of liquid flow in overlapping rings from the 120° spray angle nozzles can never distribute reflux flow above a bed of structured packing to efficiently and effectively contact the "Up-Flowing Vapors"! This compromises fractionation efficiency.

It was discovered about 55 years ago that the use of structured packing in fractionator service would not produce a reasonable HETP (Height Equivalent to a Theoretical Plate) unless a gravity-type distributor (as shown in Figure 17.3) to distribute the reflux flow above the top of the packed bed was used. The leading tower internal manufacturers have developed really excellent, and effective reflux liquid distributors, similar to what I have shown in Figure 17.3.

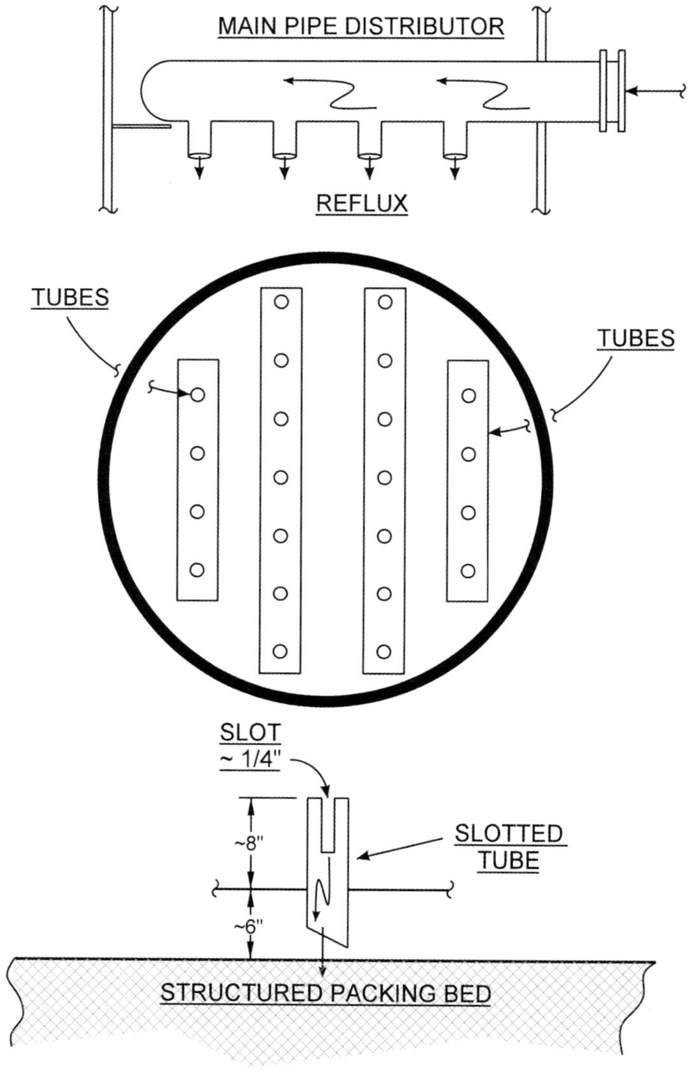

Figure 17.3. Packed tower feed distributor

Height Equivalent to a Theoretical Tray or Stage or Plate (H.E.T.P.)

The definition of a theoretical stage is that "the vapor leaving the stage must be in equilibrium with the liquid leaving the stage." This requirement cannot be obtained with a spray nozzle distributor. Thus, the need for the complex "Gravity-Type" distributor detailed in Figure 17.3.

If both the vapor flow to the bed of structured packing and the reflux liquid are very evenly distributed, then the HEPT for the packed bed may be in the range of 30" to 40". I have seen this excellent performance only a few times in my long career. Provided that the three levels of the internal reflux distribution detailed in Figure 17.3 are all completely installed CORRECTLY! That is:

1. The main pipe distributor
2. The rectangular riser distributor
3. The slotted tube distributor

Pressure Drop & Capacity

The two main advantages of using structured packing, rather than trays, are:

1. Lower pressure drop, perhaps less than half, compared to trays for an equivalent HETP.
2. Greater capacity, perhaps maximum 50% to 70% higher vapor capacity, as compared to trays.

In a refinery vacuum tower, the lower DP of the structured packing reduces the tower's flash zone pressure, which greatly increases the gas oil recovery from the asphalt bottoms.

What Then Is the Problem?

The problem based on my experience with structured packing is related to:

1. Improper installation of the liquid reflux distributor.
2. Deterioration of the liquid distribution over time.

On many, if not most, occasions, even if the initial liquid reflux distribution and the fractionation are per design on start-up, the observed fractionation efficiency eventually declines, typically, over a period from six months to two years. On one amine fractionator that I was hired to troubleshoot, this happened over a period of over 12 years.

But, only too often, fractionation is poor on start-up. Why?

It has to do with the complexity of the installation of the liquid distributor. Referring to Figure 17.3, each of the three distribution layers has to be installed so that they are completely LEVEL! But, once the top pipe distributor is installed (note that Figure 17.3 does not show the protruding parallel arms that interfere with access), the two lower levels cannot be checked by the unit engineer or the inspector for installation levelness and/or for integrity.

I once solved a problem of very poor fractionation efficiency for this sort of installation by:

1. Shutting down.

2. Removing the distributor entirely.
3. Telling the workers that I will treat them to an expensive dinner if the tower performance improves on start-up, to design specifications.

At least on this one occasion in Shreveport, Louisiana, this technique worked on a lube oil vacuum fractionator.

Reflux Maldistribution

The leading cause of distillation problems in the refinery industry is poor performance of distillation towers, using the newest forms of structured, type packing, instead of the old-style distillation trays. Figure 17.4 illustrates the problem: that is, poor distribution of reflux due to installation (not design) problems of the internal reflux distributor.

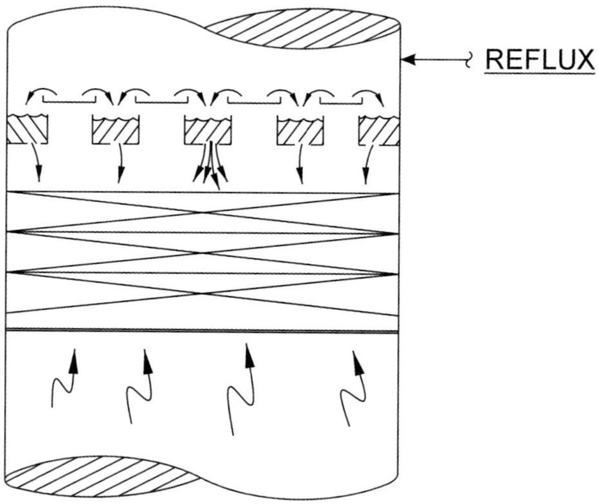

Figure 17.4. Poor reflux distribution

Often, the initial installation works quite well. But over time (months or years), the reflux distribution deteriorates, as shown in Figure 17.4. It happens in 50% of the installations I have worked on. Exactly why, I don't know.

But what I do know is that to provide my clients with a replacement design for the beds of structured packing, my response is almost always the same: TRAYS. A tray installation is easy to inspect for levelness and for proper assembly by removing each tray deck inspection manway.

It's true that I sacrifice capacity and fractionation efficiency by retrofitting a distillation tower from beds of structured packing to fixed valve cap trays. But that's the price I had to pay, to be sure to obtain the desired fractionation efficiency, in a recent project between low-valued naphtha and the much more valuable kerosene (i.e., jet fuel) products.

Summary

The use of beds of structured packing, as an alternative to trays, will certainly increase capacity and, in theory, improve fractionation efficiency. However, in practice, unless the internal reflux distribution is good, fractionation with a packed bed, for an equivalent height of packing vs. trays, may be poor. The problem is not with the internal reflux distributor design, but the distributor installation and its inspection. This is a major industry-wide problem, which I experience frequently.

18

PACKED TOWER PROBLEMS

I have a long experience in the design, operation, but mainly troubleshooting, of packed towers in hydrocarbon distillation services. Most of it was bad!

Packing is fine for pumparounds or heat removal services. I have used beds of structured packing, or grid, in crude distillation and coker fractionators many times, with good results. But in general, packed towers in fractionation services tend to be a problem. This is true for traditional dumped-type packings, such as:

- Raschig rings
- Saddles
- Pall rings
- Crushed beer bottles

as well as the more recent types of packed beds constructed from:

- Structured packing
- Layers of grid

and for new designs, in the following services:

- Absorbing
- Stripping
- Distillation

It's much better to stay with a well-designed distillation tray, with a fixed (i.e., "non-movable") valve cap, using at least a 7 to 9 mm valve cap lift, than to risk failure with packed towers.

Theoretical Advantages of Packing in Distillation Services

In theory, the main advantage of a packed tower, compared to a distillation trayed column, is capacity. A well-designed packed tower ought to require perhaps only 65% of the cross-sectional area of a trayed tower, using valves, or other types of perforated tray decks.

In theory, the other advantage of a packed tower over trays is "HETP": Height Equivalent to a Theoretical Plate.

That is a single "step" on the "McCabe-Theil Fractionation Diagram," or a fractionation stage of 100% tray efficiency.

A typical, well-designed, distillation tray has a 70% to 80% fractionation efficiency, and is set on a 24" to 30" tray spacing. Each theoretical plate, or a theoretical fractionation stage, requires about three feet of tower height for a trayed distillation tower. A typical "HETP" for a packed bed is in the range of only two feet.

Packing Fractionation Problems

I have many stories about problems with distillation towers that relied on packing. Let's refer to Figure 18.1. The reflux flows into a pipe with smaller outlets, to distribute the reflux. This distribution pipe has to be supported by a 2" wide support ring or shelf. The most common error in refinery packed bed distillation service is clamping the reflux distributor to this support too TIGHTLY, as shown in Figure 18.1 to the support ring.

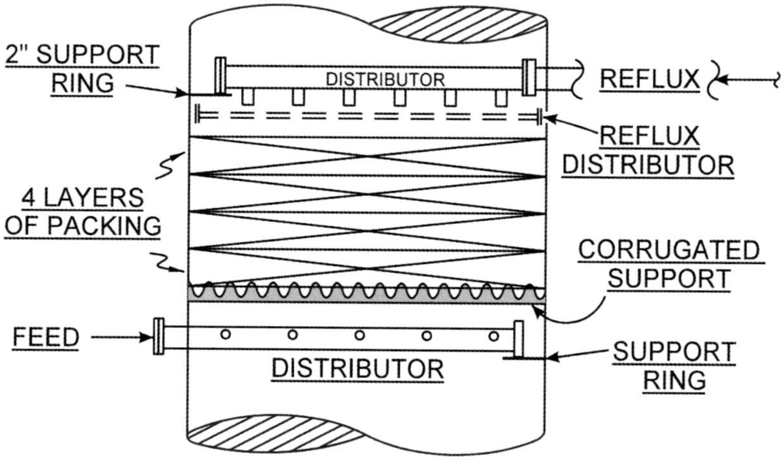

Figure 18.1. Packed tower

The differential rates of thermal expansion, between the vessel wall and the distributor pipe, are problematic here. If the end of the pipe cannot move freely as it heats and/or cools, I have seen that its flanges are forced apart, to accommodate the differential rates of thermal expansion between the vessel and the distributor pipe.

Reflux Distributor Levelness

How level must the liquid reflux distributor be (Figure 18.2), to ensure even reflux distribution through the packed bed below? Let's assume that the pressure drop through the reflux distribution holes is equal to 3" of liquid, as calculated in the following equation:

$$DH = K \cdot V^2$$

Where:

DH = pressure drop in inches of liquid,

V = Hole velocity, in feet per second,

K = The orifice coefficient (typically 0.3–0.5).

If the reflux distribution plate is one inch out-of-level (as shown in Figure 18.2), then one portion of the tower will receive a lot more reflux flow than the other portion, which is certainly bad for fractionation efficiency!

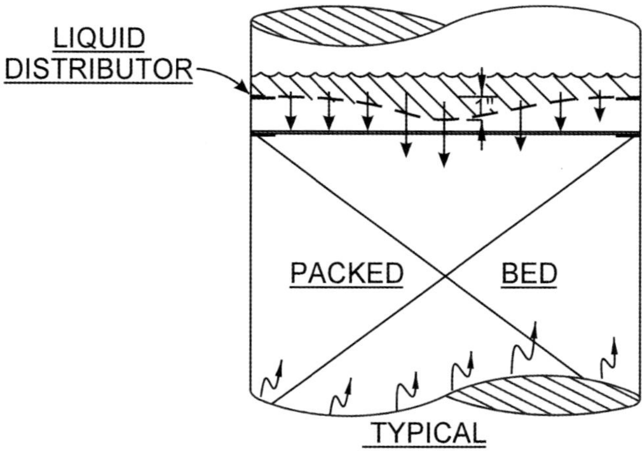

Figure 18.2. Poor liquid distribution due to an out-of-level distribution plate

Fouling

A rather small amount of corrosion deposits that are unequally distributed on the top of the packed bed will also have the same effect of causing poor reflux distribution, and hence a loss of fractionation efficiency. I minimize this problem in the wash oil sections of vacuum towers by always using an external, bucket-type DUPLEX FILTER (i.e., two filters in parallel), but without any by pass. Two filters are needed for on-stream cleaning. If a by pass is provided, some operator will use it, which is BAD!

Note that most packed bed fouling often occurs during the first day or two after start-up.

Layers of Structured Packing

Typically, each theoretical stage of structured packing might consist of two or three layers of structured packing, which is normally 10" to 11" in height per layer. Packed heights of more than 10 to 12 ft. promote poor vapor-liquid distribution, and a loss of fractionation efficiency.

The most important feature that we have to specify is the "Crimp Size" of the packing. The smaller the crimp size, the smaller the "H.E.P.T.," as defined above, and the greater the fractionation efficiency of the packing.

I have extensive experience with this problem. A really small crimp size, which provides excellent theoretical fractionation efficiency, is 0.5" or less. A fairly large crimp size, which greatly diminishes theoretical fractionation efficiency, is 1.0" or more.

Smaller crimp packing size promotes fouling, poor reflux and vapor distribution, loss of fractionation efficiency due to reflux and/or vapor channeling, and reduced fractionator run lengths.

I told this to the engineers at a refinery in Saudi Arabia, but they refused to listen and then fouled their fractionator in a few months, using seven layers of ½" crimp size structured packing. Twice in only a year!

The third time, they used, as per my direction, two layers of 1" crimp size structured packing and ran for several years between turnarounds.

Packing Support Open Area

Referring again to Figure 18.1, let's focus on the "Effective Percent Open Area" of the packing. This area is less than the open area of the packing itself. It is calculated by multiplying:

$$(\text{Open area packing}) \cdot (\text{Open area of support}) \cdot (F)$$

If the packing has an open area of 75%, and the packing support grid has an open area of 80%, then their percentage of the open area at their mutual interface would be:

$$(75\%) \cdot (80\%) = 60\%$$

But what about "F" in the above equation? That is the factor that allows for the fouling materials that will always accumulate between the bottom layer of packing and the support grid. I cannot say why this is such a common problem. But in Texas, between 1974 and 1977, when I was in charge of a number of fractionators with 2" to 3" rings of packing, I did repeatedly observe, when the tower was opened during turnaround, that:

- The rings were removed from the top-down into five-gallon buckets.

- Most of the fouling that had caused my unit to shut down due to fractionator flooding was between the bottom layer of rings and the corrugated support plate (see Figure 18.1), based on my observations as the buckets were removed from the fractionator.

To mitigate this "F" factor, I replaced the flat Packing Support Plate with a corrugated support, as shown in Figure 18.1. This did not diminish the rate of fouling accumulation on the packing support grid, but it did increase the open area of the support grid (as a percentage of the cross-sectional area of the tower), from 80% (as per above) to 120%.

Summary

In distillation services, it is best to avoid packed beds and rely on trayed towers with a fixed valve cap assembly. We should not use valve trays with movable flutter caps, as they tend to get stuck to the tray floor, because of fouled tray decks. For the fixed valve cap, I use a distance between the tray deck and the underside of the cap of:

- 8 mm: clean services
- 12 mm to 16 mm: dirty services

Perhaps the most common problem in distillation columns that use beds of structured packing is flooding due to corrosion deposits on the top of the packing.

A refinery kerosene steam stripper (Figure 18.2) was flooding at 60% of design capacity. A radiation scan (see Chapter 31) that measures the percentage absorption of radiation in the stripper at various heights showed clearly that the flooding was confined to the top few inches of the packing.

Since the vapor load was constant throughout the dual packed beds, this could only have been due to fouling deposits restricting the vapor flow through the top bed. The resulting high, localized DP of the vapor flowing through the top of the packing, resulted in liquid back-up, and hence flooding of the top packed bed.

19

STRUCTURED PACKING – PROBLEMS AND ADVANTAGES

Certainly, the very heart and soul of chemical engineering and refinery technology is distillation. About 50 years ago, in the mid 1970s, the use of structured packing was introduced in the industry. Originally, in my experience, this consisted of beds of grid to be used in the pumparound sections of vacuum towers for heat removal, but not for fractionation services.

A few years later, structured packing of a different sort was promoted by the manufacturers of trays, to be used in fractionation services. Unlike the grid, the new structured packing (see Figure 19.1) was made out of extremely thin sheets of 316, 317, or 304 stainless steel that were perforated and pressed together using extremely thin rods. This new structured packing had a large open area, as compared to the typical conventional packing of:

- Pall rings
- Raschig rings
- Saddles
- Broken beer bottles

The structured packing was intended not only to replace other types of packing, as listed above, but to entirely replace:

- Valve trays
- Sieve trays
- Bubble cap trays

and also to replace the grid-type packing introduced a few years before.

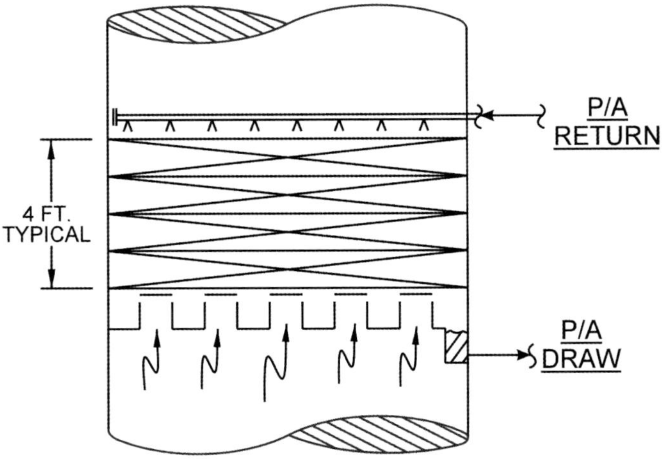

Figure 19.1. Structured packing P/A service

Reflux Distribution

For the last half-century, beds of structured packing have been preferred to trays in the refining industry. In a very large percentage of the projects that I have become involved with, the use of structured packing in fractionation services (but not in pumparound heat removal services) has been disappointing to a rather large extent.

This was not due to any problems with the structured packing itself. Its H.E.P.T. (Height Equivalent to a Theoretical Tray) is potentially the approximate 30" advertised by the tower internal vendors. And the advertised capacity of the beds of structured packing does exceed the capacity of valve trays by 50% to 60%.

The problem that plagues the use of structured packing in fractionation service is the "Reflux Distribution." It is not the design of the reflux distributor. The designs that I have seen are fine, and they can, and do, work exceedingly well, but only if these distributors are INSTALLED CORRECTLY!

The reflux distributors typically consist of three layers of distribution:

1. A pipe distributor
2. Followed by a tray distributor
3. Finally, a slotted set of small multi-tubes
   (This is discussed in detail in the following chapter.)

Once this complex installation is completed, it cannot be inspected, or tested, by the Unit Engineer. To make sure that the installation of the complex distributor is done correctly, a knowledgeable and dedicated technical person, like you or me, needs to stay in the tower for the three-, four-, five-, or eight-hour shifts that it takes to install all components of the complex reflux distributor.

A few years ago, at a lube oil plant in Shreveport, Louisiana, I was following up on the lube oil vacuum fractionator that I had designed, which could not produce the lube oil products on spec. I convinced the Plant Manager to have the internal reflux distributors removed and then be reinstalled more carefully. Prior to the reinstallation, I asked the foreman supervising the reinstallation work, as a personal favor to me, to make sure every component of the reflux distributors were installed:

1. Level – and tested with water
2. Tight – with a wrench
3. As per the vendor drawings

"And then, after start-up, we'll all go out to dinner together, to celebrate the lube oil products all meeting their required: flash point specification, end-point specification, and color specification."

And my use of applied technology did actually work, to enhance the lube oil production and quality.

I would estimate that one-third to one-half of the projects that I have been involved with in the last two decades, involving the use of structured packing in distillation services, after a few years of service, and sometimes immediately, do not achieve the desired results. The cause of the poor performance, on so many occasions, of the structured packing is not in doubt. Both expensive radiation scans, as well as my own infrared, hand-held surface temperature gun skin surveys, indicate that the cause of the poor fractionation in the beds of structured packing is poor reflux distribution!

And what then is my normal solution to this all too common problem?

Trays! I have the bed of structured packing replaced with three or four trays, with a fixed valve cap. For clean services, I use a valve cap lift of 6 mm. For fouling services, I use a valve cap of 14 mm "lift." With lift being the distance between the tray deck and the underside of the cap.

It's true this reduces potential vapor flow capacity by 30% to 40%, and increases the pressure drop by a factor of two or three. But that's the price we have to pay when we replace modern structured packing with archaic, fixed valve cap trays.

The "Bubble Effect" in Structured Packing

In 1980, I became the Tech Manager of the world's largest independently owned refinery. This was the Good Hope Refinery in Norco, Louisiana. The refinery fluid catalytic cracker fractionator (FCU) had just been retrofitted from valve trays to beds of structured packing, including the bottom, packed slurry oil pumparound section, to increase its capacity by 30%.

This bed of packing consisted of about 12 ft. of very dense structured packing, intended for very efficient vapor-liquid contacting. Note that this bed of structured packing was not in fractionation service, but in pumparound, or heat removal service.

The vapor feed to the fractionator was the fluid catalytic cracker reactor effluent vapor at 980°F. The purpose of the slurry oil pumparound was primarily to quench this highly superheated vapor stream of a million pounds per hour, from 980°F to 700°F.

The newly retrofitted tower was started up during my first week at work at the refinery. Unfortunately, the new slurry oil pumparound section flooded immediately, at an FCU feed rate of only 70% of the design rate. This was really very bad!!

The owner of the Good Hope Refinery called me into his office and said in an angry voice, "Lieberman. What's happened? Figure this out, and get back to me before you go home today. Or you're fired!"

The structured packing vendor had assumed that the pumparound section of the fractionator could be represented by a single theoretical fractionation stage in their computer simulation, even though the bed of structured packing they used was an excessive 12 feet high.

However, the published H.E.P.T. (Height Equivalent to a Theoretical Stage) was four feet. for this type of structured packing. When I re-ran the computer simulation for the fractionator, there were three theoretical stages for the pumparound:

12 ft. ÷ 4 = 3 theoretical stages

There was a huge amount of vapor produced inside the bed of structured packing, as the highly superheated vapor at 980°F contacted the 500°F slurry oil pumparound return. I called this giant increase in vapor flow inside the 12 feet of packing the "Bubble Effect." I then explained to the refinery owner that this surge of vapor flow, caused by the de-superheating of the 980°F vapor to 700°F, was a consequence of having three theoretical stages in the FCU slurry oil pumparound, rather than a single stage, as was assumed for the computer simulation.

The owner said, "What are you going to do now?!" in a very angry voice.

I had the newly retrofitted (i.e., from trays to beds of structured packing) FCU fractionator shut down and had eight feet of the structured packing removed so that the remaining height (i.e., 4 ft.) of structured packing would represent only a single theoretical fractionation stage.

When the fluid catalytic fractionator started back up a week later, the slurry oil pumparound section no longer flooded. And the world's largest fluid catalytic cracker unit ran at its design feed rate, without any problems.

The moral of this story is that the pumparound section of a fractionator cannot always be represented on a computer simulation as a single theoretical stage. This was the mistake the packing vendor made when calculating the vapor load inside the slurry oil pumparound section. Depending on the height of the packed bed, and its efficiency, the packing in the pumparound may likely develop several theoretical fractionation stages.

Structured Packing Installation

I was working in Lithuania, observing the installation of beds of structured packing in a 20-ft. I.D. vacuum tower. I observed one of the workers had just walked across a layer of structured packing and accidentally crushed the thin upper layer with his boots! To rectify this problem of crushed packing, the two installers covered the distorted layer with the next layer of 10"-high structured packing. How this affected the tower performance, I never found out!

Summary

The use of structured packing has, in theory, great advantages compared to trays, regarding both capacity and fractionation efficiency, for a given distillation tower height. But very often, due to tower internal installation problems, or deterioration of both the distributor and of the packing with use, structured packing has to be replaced with trays, in my long experience.

20

WASH OIL GRID

The key to the application of technology to improve the performance of a distillation tower is direct field observations. The incident described below illustrates this principle. It occurred in 1989, at a refinery in eastern Canada. This refinery was, at the time, 50%-owned by a major American refiner, who also supplied the technical support for the refinery.

The problem I was asked to troubleshoot was dark gas oil from the vacuum tower. The gas oil was dark because of the entrained tar rising from the flash zone (i.e., the feed zone) of the vacuum tower. The entrainment was supposed to be removed by the wash oil spray above the bed of structured packing (Figure 20.1).

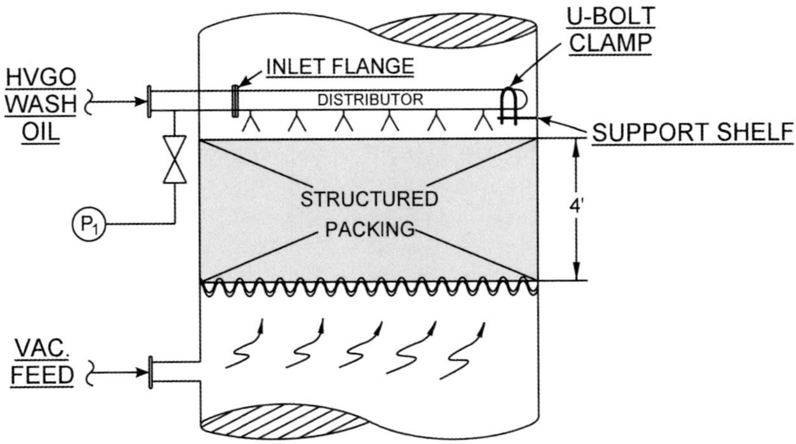

Figure 20.1. Tight "U"-bolt clamp causes thermal stress to distort Wash Oil Distributor

My client had worked on this problem for about ten years and failed to solve it. I worked on the problem for one day and arrived at an effective solution. The difference between my approach to problem-solving and my client's is something I learned from my mother as a child growing up in New York City. "Norman," my mother would say, "rely on yourself. Don't trust strangers."

Referring again to Figure 20.1, I measured the inlet pressure to the wash oil spray header (at P-1). Much to my surprise, the inlet pressure gauge read 23" of mercury vacuum. That was almost the same pressure as the vacuum tower pressure. Evidently, the pressure drop through the wash oil spray nozzles, which was supposed to be about 12 psi (i.e., 0.8 BAR), was essentially zero! Quite a bad surprise.

I next took my infrared surface temperature gun to try to determine where all the wash oil was going. Almost immediately, I discovered a cold spot (see Figure 20.1) on the vessel wall, that was immediately adjacent to the inlet nozzle of the wash oil spray header. Evidently, the internal flange (as shown in Figure 20.1) must have parted open. Thus, I concluded, almost all of the wash oil was spraying out of this defective flange and was not contacting the up-flow vapor from the flash zone (see Figure 20.1).

Past History of the Problem

But this did not make any sense to me. My client had worked on this problem for over a decade. During this time, the tower had been opened four times. Certainly, they would have inspected the wash oil distributor and repaired the leaking flange. What then was the problem? There was surely some factor that was causing the wash oil header to fail repeatedly at this flange! But what was it? Piping flanges rarely fail without some very extreme piping stresses! A very difficult problem.

Direct Field Observations

Three months passed. The management of the refinery had shut the vacuum tower down to repair the faulty flange, as they had done repeatedly over the past decade. But, on this occasion, they asked me to offer my opinion as to the cause of the repeated flange failures.

As shown in Figure 20.1, I found that the pipe distributor's far end was resting on a 4" wide shelf. To keep the end of the pipe distributor from falling off of this shelf, the end of the pipe distributor had been slipped underneath the "U" bolt clamp, as shown in Figure 20.1.

And then, to make really, really sure that the end of the pipe distributor could never, ever, accidentally fall off its 4" support shelf, this "U" bolt clamp had been tightened really, really TIGHT!!

Differential Rate of Thermal Expansion

Tightening this "U" bolt was not seen as a problem, as long as the vessel and the distributor pipe were both cold. But on start-up, the vessel would heat up to the flash zone temperature of the vacuum tower or about 720°F. The wash oil (that is the heavy vacuum gas oil pumparound return) would be much cooler, or about 350°F.

The resulting differential rate of thermal expansion would be a tremendous stress on the distributor pipe inlet flange (which was only a 4-bolt flange, not a regular, 6-bolt, 150-psi piping flange). And this thermal stress caused this inlet flange to part and permit most of the wash oil to leak out, before it could be distributed across the bed of wash oil structured packing.

Final Outcome

I loosened the "U" bolt so that it would still remain in place, but also so that the end of the distributor pipe could slide underneath it to allow for the differential rates of thermal expansion between the hot vacuum tower shell and the much cooler distributor pipe. The vacuum tower was then returned to service. The heavy vacuum gas oil quality was then excellent and remained so for many subsequent months, until the scheduled unit turnaround the following year.

Moral of This True Story

Why was I able to solve the fundamental problem in minutes, while my client had worked on the same problem for years, without success? Because I listened to my mother; I relied on myself. And that is the concept that I have relied upon for my entire career as a Refinery Process Chemical Engineer. Direct field observations are always the key to successful process problem identification and correction. They are also the key to process plant profitability.

Dangers of Pyrophoric Iron

During the 1980 strike at the world's largest refinery in Texas, I was working as an outside operator on the amine-sulfur recovery plant that I had designed two years previously. One of my duties was to replace the amine filters that became routinely fouled with $Fe(HS)_2$ (i.e., pyrophoric iron) solids.

I discarded these cartridge-type filters in a large trash bin. One day, after the spent filters had been in the bin and dried out, they began to burn and emitted a FATAL concentration of SO_2, which is just as deadly as H_2S.

The same thing can happen inside a distillation tower, especially if beds of structured packing are used. When it is opened for inspection, after a few days, the black $Fe(HS)_2$ deposits will dry out. They then will only too often catch fire, at ambient temperatures, and emit FATAL concentrations of SO_2.

A Lesson Learned

Chemically clean all distillation towers that can have iron sulfide corrosion deposits, to make sure that these distillation towers are all SAFE for entry!

The auto ignition temperature of iron sulfide (H_2S) is ambient temperatures, when the black, powdery corrosion deposits dry out.

Of course, unlike H_2S (which has no odor at fatal concentrations), sulfur dioxide (SO_2) has a very strong odor. But if you are trapped inside of a vacuum tower, this odor will not help you to escape.

Summary

Wash oil grid fouling with corrosion products in distillation towers is a common problem. Often, it is some minor installation error of the wash oil grid that causes the grid to foul prematurely. Very often, these fouling deposits will auto-ignite upon exposure to air and then release fatal concentrations of sulfur dioxide inside the distillation tower. SO_2 is just as deadly a gas, as is H_2S!

21

DISTILLATION TOWER PRESSURE PROFILE

"HEAT MAKES PRESSURE." That's the main concept that will be discussed in this chapter. "COOLING REDUCES PRESSURE" is the second concept that we will discuss. The third concept is that vapor flowing through a restriction accelerates and the energy to increase the vapor velocity comes from the pressure of the flowing vapor.

The gasoline debutanizer (Figure 21.1) shows a typical pressure and temperature profile for a light-ends hydrocarbon distillation tower. The tower top pressure floats on the pressure in the reflux drum. The pressure for:

1. Tower top (1) tray
2. Tower bottoms (25) tray
3. Feed tray (15)

All three points float on the reflux drum pressure, and vary with the vapor flow rate, through the intervening tray decks.

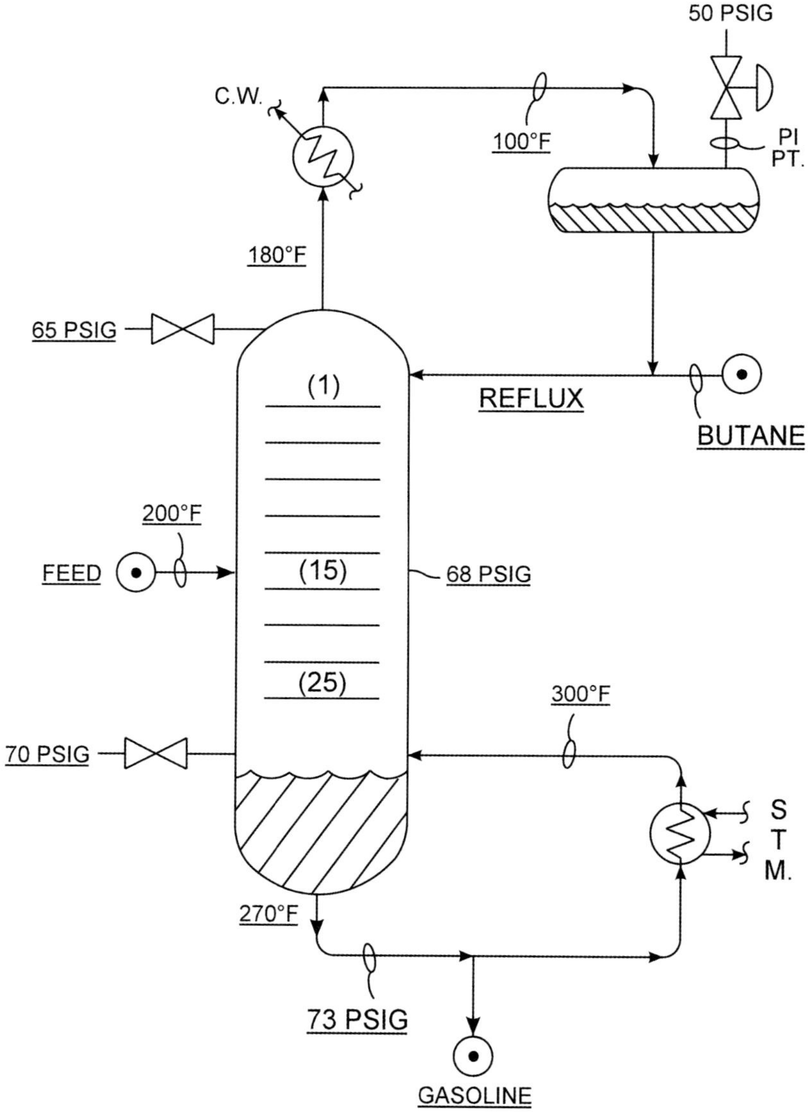

Figure 21.1. Tower Pressure Profile – De-Butanizer

Reflux Drum Pressure

As shown in Figure 21.1, the temperature in the debutanizer reflux drum is 100°F. If we had pure, 100% normal butane in the reflux drum, then the pressure in the reflux drum would be about 60 psia (45 psig at sea level). This is the vapor pressure of normal butane at 100°F. If the reflux drum was completely full of liquid (i.e., there was no vapor space) the drum pressure could be much higher. In that case, I would say the liquid in the reflux drum would be "Sub-Cooled."

If there was a vapor vent on the top of the reflux drum, then the reflux drum pressure would be controlled by the drum's off-gas, back-pressure control valve. However, in this case, it's simply the vapor pressure of normal butane at 100°F that sets the reflux drum pressure and tower pressure.

Tower Top Pressure

The pressure difference between the reflux drum and the top of the debutanizer is 15 psi (as shown in Figure 21.1). This pressure drop is due to the following six factors:

1. The condenser shell-side pressure drop
2. Vapor inlet line pressure drop
3. The debutanizer tower vapor outlet nozzle pressure drop
4. Liquid outlet-line pressure drop
5. The condenser liquid outlet nozzle pressure drop
6. The liquid line to the reflux drum pressure drop

Very often, in my experience, it is item #3 above that contributes the largest pressure drop. This DP is equal to:

$$\text{Delta P} = 0.3 \bullet (D_V \div D_L) \bullet (V)^2$$

Where:

Delta P = Pressure drop, inches of liquid

D_V = Density of vapor, lbs./ft.^{3}

D_L = Density of liquid, lbs./ft.^{3}

V = Nozzle exit velocity, ft. per second

The "0.3" factor, is a typical nozzle pressure drop, orifice coefficient.

Reflux Drum Elevation

If the reflux drum is elevated above the condenser, so as to provide "Net Positive Suction Head" (NPSH) to the reflux pump, that will also contribute to the debutanizer top pressure, as the liquid has to flow up into the elevated reflux drum.

Note that as the condenser effluent rises into the elevated reflux drum, it is not allowed to partly vaporize. But why?

Because there is NO vapor vent from the reflux drum. This problem is the leading cause of FLARING in the hydrocarbon process industry, unless the reflux drum can be vented to fuel gas.

Conclusion: It's best to elevate the reflux condenser just a few feet above the reflux drum, which was standard industry practice prior to the 1940s.

Tray Pressure Drop

The pressure difference between the top of the tower and the vapor space above the reboiler return nozzle (see Figure 21.1), is the pressure drop of the vapor flowing through the distillation tray decks.

The pressure drop through the tray decks is a result of two factors:

1. The weight of liquid on the trays
2. The acceleration of the vapor flow through the holes (or valve caps)

The weight of liquid on the trays is a result of:

1. The height of the weir
2. The crest height of liquid that is overflowing the edge of the weir (Figure 21.2)

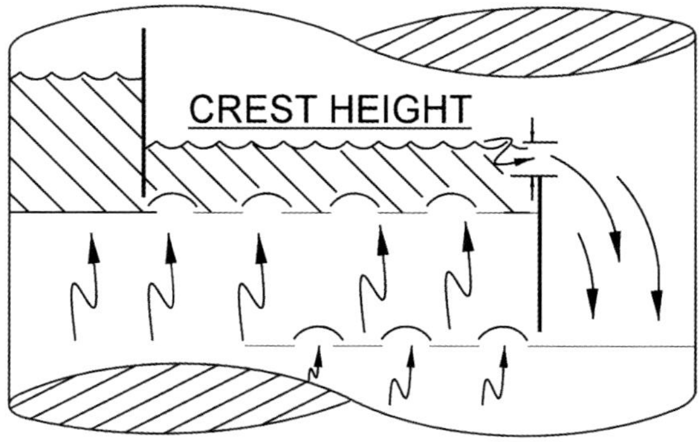

Figure 21.2. Crest height is typically 1"-1 ½" above the top of the weir

Acceleration of the vapor through the tray's perforations produces a pressure drop great enough to support the weight of the liquid on the tray decks. That is, when the vapor flows up through the tray decks and speeds up by a factor of roughly ten times, some of the pressure of the vapor is converted into the vapor's kinetic energy.

The purpose of having high-velocity vapor rushing up through the liquid level on the trays (due to the weirs) is to mix the vapor and the liquid efficiently together, which is the actual function of trays.

The better the vapor and liquid are mixed together on the tray deck, behind the overflow weirs, the better the separation efficiency between gasoline and butane, and the better the performance of the distillation tower.

Tower Bottoms Pressure

The tower bottoms pressure is then the sum of:

- The tower top pressure

Plus

- The sum of pressure drop through each of the tray decks shown in Figure 21.1

The pressure drop of the vapor flowing through the trays is a function of the heat input to the reboiler and the tray design.

Reboiler Pressure

The pressure drop of the shell-side of the reboiler, shown in Figure 21.1, is the sum of:

- The liquid nozzle entrance loss
- The feed line pressure drop
- The shell-side pressure drop, as the reboiler feed is partly vaporized, and as it flows through the shell-side of the exchanger, between the tube-support baffles
- The reboiler vapor return line pressure drop

Bottoms Liquid Level

Note that the debutanizer's bottoms level is equal (in hydrostatic liquid head) to the sum of the factors causing reboiler shell-side pressure drop (items #1 to #4, that have been tabulated above).

However, there is a further complication that needs to be included in understanding the bottoms liquid level, and that additional complication is FOAM! To some extent, the effluent from the reboiler is a vapor-liquid mixture, or foam. Depending on the surfactants in the bottoms product, this foam may break down to vapor and liquid in as little as ten seconds or as much as five minutes. However, if the level of foam rises above the reboiler return nozzle, the tower will flood due to entrainment of foam. This, unfortunately, is an extremely common problem in many hydrocarbon and refinery distillation services. Stable foam tends to form in liquids that have a high surface tension – cold water being one such example.

Summary

Very often, performing a local, field pressure drop survey of a distillation tower will reveal the fundamental problems with the installation, or design, of the tower internals. This tool is most valuable when the engineer or operator responsible for solving the problem makes the pressure measurements himself or herself with his or her own hands. Most of the successes I've had as a refinery troubleshooting consultant are a result of observations I accidentally made while conducting field pressure drop surveys.

22

REFLUX AND REBOILER HEAT FLOWS

The main concept of distillation is to adjust two factors—reflux flow and reboiler heat. The reflux removes heat from the top of a distillation tower. The reboiler adds heat to the bottom of a distillation tower. Both factors increase vapor-liquid traffic on the trays, which make the trays work harder.

If the reflux rate is increased, then the reboiler duty must also be increased. If the reflux rate is increased, and the reboiler duty is not proportionally also increased, then the reflux drum will begin to go empty. In other words, "The reflux comes from the reboiler's heat."

If we put a BTU into the bottom of the distillation tower, that BTU vaporizes about 0.01 lbs. of hydrocarbon liquid. The heat, in the form of hydrocarbon vapors, flows up the distillation tower, through the overhead condenser. And the condensed vapor then flows as liquid into the reflux drum, to replace the liquid that has, a few moments ago, flowed down the distillation tower in the form of reflux (Figure 22.1).

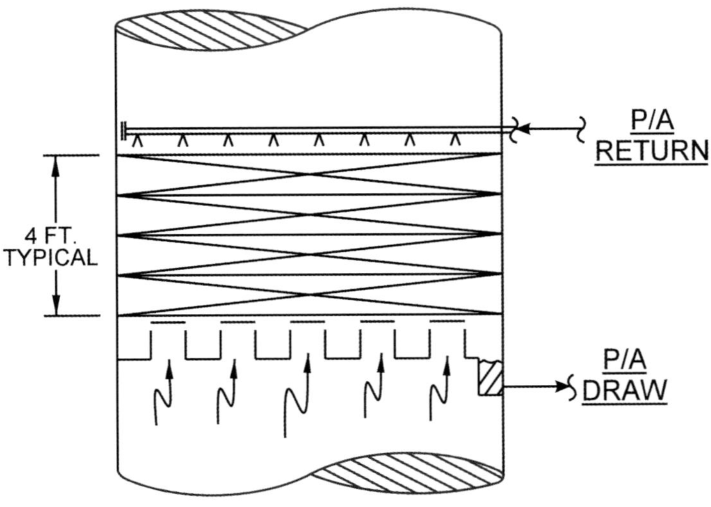

Figure 22.1. Distillation tower

And the liquid from the tower top condenser in the form of reflux, cascading counter-currently down through the distillation tower against the rising vapors from the reboiler is what separates:

- Ethanol from water,

or

- Propane from butane,

or

- Kerosene from diesel oil,

or

- Naphtha from waste water,

or

- Gas oil from vacuum residue,

or

- Most commonly, and most importantly, diesel oil from gas oil, which is the most common factor for increasing a refinery's profitability.

The most important rule to always remember about distillation is that the reflux flow comes from the reboiler. However, like all rules, there are exceptions, as I will describe below.

Effect of Feed Preheat

Referring again to Figure 22.1 (a propane-butane distillation tower), the heat needed to generate the reflux may come from two sources: reboiler steam or feed pre-heater steam. However, if some of the reflux comes from the feed pre-heater exchanger, then the reflux can be increased, but without increasing the heat input into the reboiler.

Therefore, the top part of the distillation tower will work harder as the top reflux flow is increased. But the lower part of the distillation tower does not work harder as the top reflux rate is increased.

To summarize, increasing the reflux rate will normally improve fractionation efficiency. However, if the incremental heat needed to keep the reflux drum full comes only from the feed preheater, then fractionation efficiency will usually only improve by a small amount.

But if the incremental heat needed to keep the reflux drum full comes from the reboiler, then the fractionation efficiency will improve by a larger amount, because both the top part and the bottom part of the distillation tower are working harder.

Entrainment or Vapor Flood

However, if the extra vapor flow required to keep the reflux drum full causes droplets of liquid to be entrained from a lower portion of the distillation tower to the upper portion of the distillation tower, then a reduction of the vapor flow, with either less reboiler duty, or with less feed preheat, will increase the fractionation efficiency between the propane and the butane.

The problem of loss of fractionation efficiency due to droplets of liquid being entrained is called "Flooding," which is the single largest problem we have in applying distillation technology, and is typically the limiting factor in most process plants and refineries.

Temperature Difference

To determine if more reboiler heat, or more feed preheat, is making fractionation between propane and butane better or worse, always look at the temperature difference between the top and bottom of the distillation tower.

If the temperature difference is increasing, then the distillation tower's separation efficiency is getting better. We do not have to wait for laboratory sample results to draw this important conclusion as to whether more reflux and more reboiler heat is improving the split between propane and butane products.

If you worked with me when I am troubleshooting a distillation problem, you would notice that in the short term I always rely upon the tower's temperature profile to decide if an operational move I had just made is hurting or is helping the separation efficiency. I am much too impatient to wait for lab results, but the tower temperature profile is usually readily available.

Fractionation in Condensers and Reboilers

Not all the fractionation in a distillation tower takes place in the tower itself. As I discuss in the following chapters, a very large percentage of the separation between the light and heavy components may take place in both the overhead condenser and bottoms reboiler.

If the overhead condenser is a partial condenser, and thus the overhead product from the distillation tower is only the vapor stream from the reflux drum, then a very large percentage of the tower's fractionation will take place in the reflux drum. In this case, all of the liquid condensed is used for reflux.

If the bottoms reboiler is a once-thru, thermosyphon reboiler, then a very large percentage of the tower's fractionation will take place in the reboiler, but not if the reboiler is a circulating-type reboiler. This is also discussed in great detail in a subsequent chapter of this text.

Summary

A significant amount of fractionation in a hydrocarbon distillation tower often takes place in the overhead condensers and in the bottoms reboilers. When we add or subtract heat from a theoretical fractionation stage in a distillation tower, it provides much more separation efficiency than any distillation tower tray. This is evident if we look at the output from a distillation tower computer simulation, as indicated by the ratio of the key components tabulated in the simulation output. Most especially, in a once-thru thermosyphon type reboiler, as discussed in the following chapter.

23

THERMOSYPHON CIRCULATION REBOILERS AND STEAM STRIPPING

Figure 23.1. shows a reboiler in thermosyphon circulation service. Most of our distillation tower reboilers circulate by this method, rather than the less common type of forced circulation using a centrifugal pump.

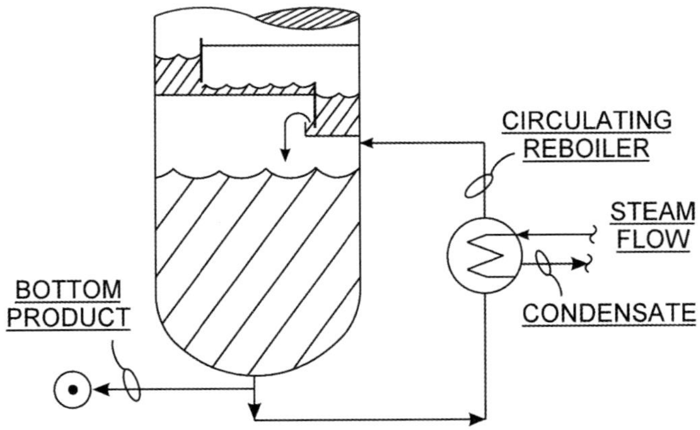

Figure 23.1. Thermosyphon circulation

The feed flows into the reboiler due to the liquid head in the bottom of the tower. The outlet from the reboiler has been partly vaporized by the reboiler's steam flow. This vaporization reduces the mixed-phase density of the effluent from the reboiler flowing back to the tower. This reduction in density in the reboiler return line reduces the outlet pressure of the reboiler, as compared to the reboiler inlet pressure.

This difference in the pressure between the reboiler's inlet line and the outlet line is what drives the reboiler circulation or flow. Increasing the steam flow to the reboiler, increases the vaporization, reduces outlet density, and will increase the thermosyphon flow circulation rate, through the shell-side of the reboiler.

Circulating Reboiler

The reboiler shown in Figure 23.1 is called a "Circulating Thermosyphon" reboiler, because the feed to the reboiler coming from the bottom of the distillation tower is partially vaporized and then both the vapor and liquid return to the bottom of the tower.

Some of the returned liquid flows out of the bottom of the tower, as the bottoms product. But, usually, most of the returned liquid is recirculated to the reboiler. This recirculation of the reboiler effluent reduces the fractionation efficiency of the reboiler, oftentimes by a lot, which is bad.

Once-Thru Reboilers

Many process unit reboilers used in distillation services are called "Once-Thru" reboilers (Figure 23.2), because the feed to the reboiler is only flowing from the bottom tray of the distillation tower. None of the flow to the reboiler comes from the bottom of the tower. In the once-thru reboiler:

- All the flow from the bottom tray flows through the reboiler.
- The bottoms product consists of only the liquid portion of the reboiler effluent.

Once-thru reboilers contribute a lot to the fractionation efficiency of a distillation tower, which is good!

Circulating reboilers contribute very little to the fractionation efficiency of a distillation tower. Which is bad!

If the percent vaporization of the feed in a once-thru reboiler exceeds 50% to 60%, we are supposed to use a circulating reboiler to prevent an excessive amount of vaporization. Otherwise, using a once-thru reboiler is preferred and will greatly improve fractionation efficiency.

I have discussed this very critical subject in greater detail in the following chapter. In my very long experience, converting a circulating-type reboiler to a once-thru reboiler, is really one of the most effective changes we can make, to improve the fractionation efficiency of any distillation tower.

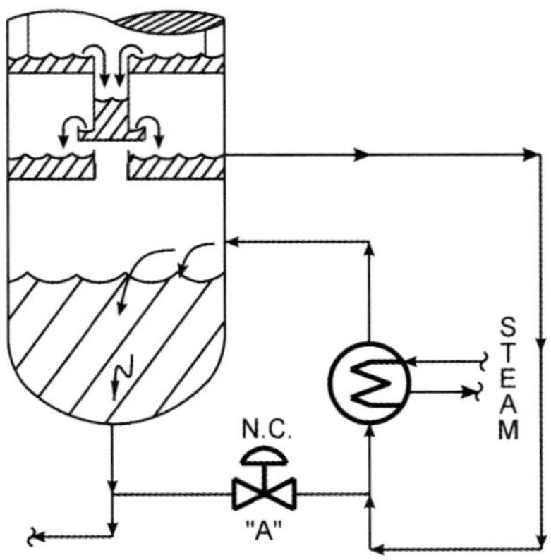

Figure 23.2. Once-thru thermosyphon reboiler

Reboiler Pressure Drop

As the percentage of vaporization in the reboiler shown in Figure 23.1 increases, the rate of circulation should increase as well, which is good, as it keeps the reboiler cleaner due to the higher flow rate of the liquid. However, as the percentage of vaporization increases, the pressure drop through the reboiler circuit (especially through the shell-side of the reboiler) will increase. This increase in shell-side pressure drop will tend to choke-off the shell side flow, which is bad, as it promotes shell-side reboiler fouling.

Function of the Reboiler

Why do almost all our distillation towers have reboilers? For three reasons:

1. To vaporize some of the lighter material in the feed into the overhead product.
2. To heat the bottoms product from the feed temperature to the higher distillation tower bottom's temperature.
3. But mainly, most of the reboiler heat is used to generate the distillation tower top REFLUX flow.

The incremental reflux flow, often needed to improve fractionator efficiency in a distillation tower, originates in the reboiler. THAT THE REFLUX COMES FROM THE REBOILER is the key fact to remember, when trying to enhance a distillation tower's fractionation efficiency. It is actually the key concept in all distillation technology.

Improving Reboiler Capacity

One of the most important, but surprising, aspects of the heat transfer efficiency of a thermosyphon reboiler (i.e., the heat transfer coefficient) is the smoothness on the outside tube surface area used in the reboiler, as shown in Figure 23.1. The incident I'll now describe occurred in a refinery in Texas City, in 1969, where I was engaged in field troubleshooting as a very young process engineer.

A True Story

I had an old iso-pentane reboiler using 30 psig steam, which I decided to replace with a new reboiler bundle. The old tube's exterior surface area was rough and pitted. The new tube bundle's exterior tube surface area was bright, smooth, and shiny. However, much to my dismay, when I started back up with the new, smooth tube reboiler bundle, my overhead reflux drum went empty.

Bobby, my friend and my oldest operator, suggested I put the older, rough surfaced reboiler bundle back into service. I decided to follow his advice. A few hours later, the de-pentanizer reflux rate was fully restored. Why did the rough-surfaced, old tubes work so much better than the new, smooth tubes?

For heat transfer coefficients involving a pure component (iso-pentane), with the boiling fluid on the shell-side (as in Figure 23.1), a rough, pitted surface may develop three times the overall heat transfer coefficient (i.e., "U" value), than would a tube with a smooth, shiny surface. This principle would only apply to cases in which the majority of the heat to be transferred from the steam is used for latent (i.e., boiling) heat, rather than for sensible heat (i.e., the increase in temperature) duty.

Sand blasting the outside of a new tube bundle, in latent heat transfer service, is an effective way to use this principle to increase reboiler capacity. This will roughen the surface so as to provide sites where the nucleate boiling can be initiated.

Steam Strippers

In a steam stripper (in hydrocarbon service), the steam does not provide the heat that would normally be provided by the reboiler. The heat of vaporization in a steam stripper does not come from the steam. Rather, it comes from the sensible heat content of the feed itself.

For every 25°F temperature reduction of the feed as it flows through the steam stripper, 15 wt.% of the feed is vaporized:

25°F temperature drop = 15 wt.% vaporization

The function of the steam is not to supply this heat of vaporization. It is to reduce the "Hydrocarbon Partial Pressure" of the feed. It is this reduction in partial pressure that promotes the vaporization in "Refinery Gas Oil Side-Stream Stripping Towers"!

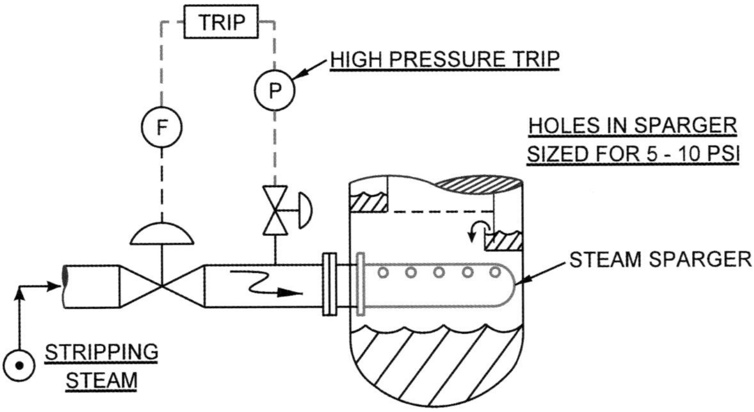

Figure 23.3. Steam stripper

Moisture in Stripping Steam

While the stripping steam does not provide any heat to vaporize the stripper feed, it very often greatly diminishes the stripper efficiency, if it contains liquid water. Water in the stripping steam is vaporized when it mixes with the much hotter oil on the stripper's tray decks. My field tests have shown that cooling the oil by 40°F, by either:

1. Water in the stripping steam,

or

2. Ambient heat loss from the stripper itself, or the stripper's feed line,

which reduces the stripper operating temperature by more than 40°F to 50°F, will reduce the stripper efficiency to ZERO!

In refineries that I have worked in recently, these two malfunctions have substantially reduced the production of the refinery's most valuable product (i.e., jet fuel or kerosene) and greatly reduced plant profits.

Steam Stripping Efficiency

For every 25°F temperature drop of the stripper feed due to the flowing steam, about 15 wt.% of the feed is vaporized. This assumes:

- Latent heat of vaporization of feed = 100 BTU/LB
- Specific heat of feed = 0.65 BTU/LB/°F

This does not include any ambient heat losses from:

- Stripper feed line
- Stripper vessel itself

And this does not include any allowance for water evaporation from wet steam.

Once the sum of these three items listed above exceeds 40°F to 50°F, stripping efficiency will be zero. This is quite a common problem in a refinery, in my 60 years of experience, and very often greatly reduces refinery profitability.

Summary

The accidental conversion, in a refinery distillation tower, of a once-thru thermosyphon reboiler to a circulating thermosyphon reboiler by opening up the start-up valve and connecting the bottom of the distillation tower directly to the reboiler inlet, is one of the most common errors in applied distillation technology.

Lack of optimization of steam stripper tray hole areas on the individual stripping trays is an almost universal design error. The hole area on the bottom stripping tray has to be far less than on the top stripping tray, by roughly a factor of three-to-one, after taking into account the increase in molecular weight of the vapor flow to the upper trays, as compared to the lower trays of the steam stripper, and the hydrocarbons that are vaporized by the stripping steam flow. The hole areas on the stripping trays should be selected so that the tray vapor pressure drop is about the same for all of the trays.

24

ONCE-THRU THERMOSYPHON CIRCULATION REBOILERS – CASE HISTORY

In June 1975, I was in charge of the world's largest alkylation unit in Texas City. My alky unit produced 26,000 BSD of high-octane alkylate. The 100-octane alkylate product flowed from the bottom of the de-iso-butanizer tower. The overhead product was isobutane that was recycled to the alky reactor.

Normally, the isobutane content of the alkylate was 4%. However, when I came to work one morning, the on line gas chromatograph showed that the isobutane content of the alkylate was not its normal 4%, but 8%. This meant that the "Reid Vapor Pressure" of alkylate was much too high, because its isobutane content was excessive.

Other than the above catastrophic situation, tower operation was normal. It was catastrophic because the alkylate was flowing into a floating roof storage tank. If high vapor pressure alkylate flowed into the tank for several hours, the floating roof might sink on the world's largest alkylate storage tank, which would be a rather negative development in my engineering career.

Super Fractionation Stage

A night shift operator had accidentally opened valve "A" (shown in Figure 24.1). This allowed a portion of the liquid from the bottom tray to by-pass the reboiler, and drop directly into the tower's bottom product. The reboiler functions as a "Super Fractionation Stage"! Partly by-passing the reboiler with the bottom product, as shown in Figure 24.1, GREATLY reduces fractionation efficiency. This substantially increases the amount of I-C_4's in the alkylate, from its normal maximum of 4%, to the excessive 8%. (Note that between 5% to 8% of the gasoline sold in the U.S.A. is normally alkylate.)

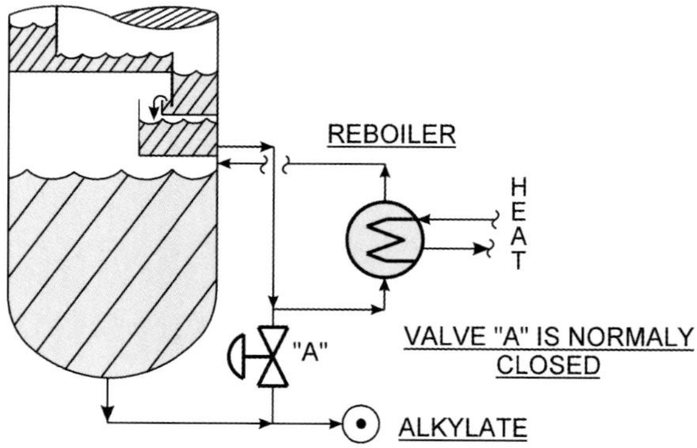

Figure 24.1. Valve A: Keep closed until reboiler circulation is established

The Function of Valve "A"

Valve "A" is intended to allow liquid from the bottom of the tower to flow directly into the reboiler on start-up, when there is very little liquid flow going into the reboiler, because most of the liquid from the bottom tray is leaking through the bottom tray deck.

But why?

Because without vapor flow through the bottom tray, there is nothing to stop this tray from leaking, as it is a valve-type tray and not an old-style bubble cap tray.

Valve "A" allows liquid from the bottom of the tower, which is leaking through the bottom tray on start-up, to gravitate into the reboiler, until enough reboiler vapor is generated to stop the bottom tray from dumping liquid directly into the bottom of the tower and into the tower's bottom product. This flow by passes the reboiler and thus reduces the vapor flow to bottom tray. Note that opening valve "A" will only increase flow into the reboiler, if the bottom's liquid level is at, or above, the reboiler's lower level. A mistake I have made several times.

Restoring Fractionation Efficiency

With valve "A" open, the bottoms product, rich in isobutane, will partially by-pass the reboiler. This allows the isobutane content of the alkylate bottoms product to increase from 4% to an intolerable 8%!

I then closed valve "A" and fractionation efficiency was restored.

Addition of heat to a fractionation stage below the feed inlet greatly enhances the fractionation efficiency of that stage. Opening up the start-up valve by passes this "Super Fractionation Stage."

Just like all employees are not of equal value to the refinery, all fractionation stages in a distillation tower are not of equal value to achieving product specifications.

If you study a tray-to-tray computer simulation output, you will observe that the ratio of the key components changes far more rapidly on the bottom theoretical stage, which is the "Reboiler," than on any other of the fractionation stages in the tower below the feed tray.

Origin of Start-Up Line

I started work for American Oil in 1965 at their giant refinery (400,000 BSD) in Whiting, Indiana. This was about the time that many fractionation towers were being retrofitted with valve trays, to replace old-style bubble cap trays. The valve trays have about 10% to 15% more capacity than do the bubble cap trays.

However, on start-up, much to the surprise of the operational personnel, they could not establish proper liquid circulation of the bottoms product into the reboiler, and hence could not generate the required reflux rate for normal fractionation. To correct this problem, a jump-over line was installed to feed the reboiler, on start-up, from the tower bottom's liquid level.

Bottom Tray Leakage

Of course, if the bottom tray has been damaged, due to running with a tower bottoms high liquid level (a very common mistake), then leaving the start-up valve (i.e., valve A shown in Figure 24.1) open is necessary to establish an adequate flow of reflux. This is an unfortunate, but all too common, problem on many of the distillation towers I have worked with.

Summary

Leaving the "Start-Up Valve" accidentally open on a distillation tower, once-thru thermosyphon reboiler, is a common operational error. This problem may be entirely avoided by using a "Total Trap Out Chimney Tray" to feed the reboiler, thus entirely avoiding the need for the start-up valve. Also, the chimney tray avoids the loss of fractionation efficiency due to possible leakage from the bottom fractionation tray deck into the bottom of the tower.

Leaving Valve "A" Open

A fellow chemical engineer asked me last week, "What is the most common error in distillation tower operations?"

My answer was, "Leaving valve "A" (Figure 24.1) open after start-up."

Of course, if the bottom tray is subject to severe tray deck leakage, then valve "A" has to be left open all the time. To avoid this problem, I use a total trap-out chimney tray to feed the reboiler, which does not need any vapor flow to avoid leakage of liquid through the tray deck. Better to be safe than sorry.

When I worked in Texas City, I noticed that when they started up a fractionator, my operators always left valve "A" open. Then, a few hours later, they would close it. I asked them why they followed this procedure? "Mr. Norm," they responded, "the tower works better with valve "A" closed after start-up!!!"

And so it does!

25

REBOILER STEAM QUALITY

Moisture in steam can reduce the capacity of a distillation tower reboiler. The origin of this detrimental moisture is commonly the carry-over from a boiler steam drum. Not only is moisture in the steam in itself bad, but the droplets of entrained water will contain hardness deposits. These hardness deposits (typically silicates) may precipitate inside of the reboiler's tubes.

The hardness deposits in the boiler feed water supply are measured in units of TDS (Total Dissolved Solids) in parts per million (PPM).

Normally, we blow down 5% to 10% of make-up BFW to purge out the TDS (hardness deposits or the silicates). If you look at your boiler's steam drum, you will see that the blowdown valve is NOT located (see Figure 25.1) at point "2" (i.e., the bottom), but at point "1"—the upper side.

But why?

Because the TDS tend to accumulate at the steam-water interface and not settle out at the very bottom of the boiler.

Level Control

Since the solids tend to concentrate at the steam-water interface in the boiler steam drum, small amounts of carry-over due to high BFW levels will result in high concentrations of hardness deposits in the generated steam. This is why the boiler blowdown is not at the bottom of the boiler, but at the higher steam-water interface.

But how do we know where this interface level is? A difficult problem!

The normal, but wrong, answer is that we could look in the level gauge glass, or the output from the LRC (Level Recorder Indicator). However (and this is the main point of this chapter), the level-glass and the LRC output are **always <u>WRONG</u>!** Wrong, because they always will show a level that is lower than the actual level in the boiler.

But why?

The level glass is always somewhat cooler than the boiling water in the steam-water separator as depicted in Figure 25.1. For every 60°C (100°F) difference, the colder water contracts by 2% to 3%, which is a small problem that we can neglect.

However, boiling water is full of bubbles of steam, which reduces its mixed-phase density by somewhere between 10% to 50%, depending on the foaming tendency of the boiler feed water. The foaming tendency is a function of the concentration of surfactants in the boiler feed water, which reduces the surface tension of the boiling water. A reduction in the surface tension of any liquid is the origin of "Foaming."

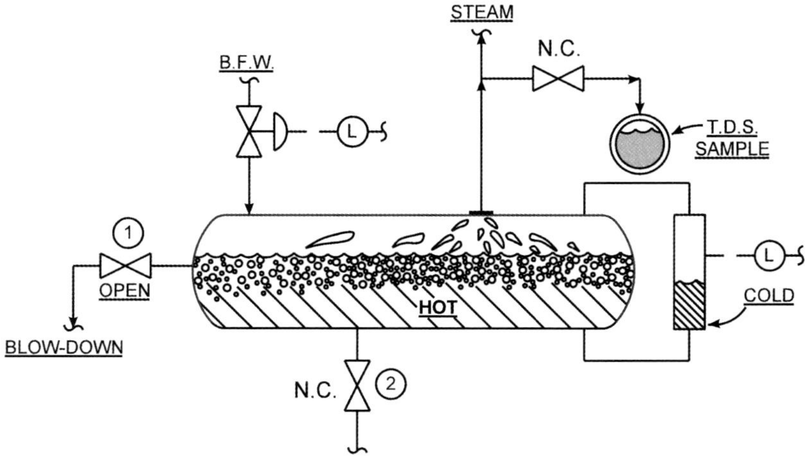

Figure 25.1. Gauge glass level is lower than steam drum boiling level

Internal vs. External Boiler Level Measurement

There are two measurements then that can be used to correct for the foaming tendency of the boiling water in the steam-water separation drum (i.e., the STEAM DRUM), shown in Figure 25.1. The industry-accepted method practiced by the chemical vendors that sell chemical additives is to sample (as shown in Figure 25.1) the Total Dissolved Solids (TDS), from a sample of the condensed steam. A TDS probe costs $10 at Walmart, and the sample of the condensed steam only takes a few minutes to obtain.

But then, there is my way to check for excessive levels of foam. Whether the heating medium is hot flue gas in the convective section of a fired boiler, or circulating hot oil from a fractionator pumparound, I will proceed as follows:

1. Lower the indicated external water level by about an inch. That is the level that you see in the gauge glass.

2. Wait 5 to 10 minutes. The steam effluent temperature will usually not change, because the upper row of tubes are still submerged in the boiler feed water, and thus they are still promoting poor quality steam, if the water level is still too high.
3. Lower the indicated external water level by another inch and wait 5 to 10 minutes. If the evolved steam from the boiler does not increase in temperature, keep slowly reducing the level in the steam drum.
4. Eventually, the convective section effluent flue gas temperature from the boiler will get a little hotter. Or, if you are generating the steam from a waste heat kettle-type steam generator (with hot gas oil pumparound on the tube side), the hot oil pumparound effluent from the horizontal tube bundle will get a few degrees hotter. (This is my favorite method in refinery pumparound steam generation coolers.)
5. You have just uncovered the top row of tubes in your boiler. Or you may have lowered the water level so much as to have excessively reduced the water's rate of thermosyphon circulation, and thus inhibited heat transfer in the boiler's tubes.
6. Steam production may also diminish.

I have not noted any effect on the evolved steam temperature using the above technique, only an increase in the effluent temperature of the heating medium.

On kettle-type boilers, with a hot oil pumparound loop on the tube-side, I have typically observed that the water level in the level gauge glass will be typically 8" to 12" below the top row of the tubes inside the kettle itself. That difference is largely a function of the tendency of the water to foam in the kettle, and thus of the boiling water's density to diminish.

Silicates

My main interest in reducing the moisture content of steam is to keep the silicates (i.e., hardness deposits) out of the steam. These silicates may contribute to the tube-side fouling of our distillation tower reboilers.

Also, if the distillation tower is a vacuum tower, with vacuum steam jets, the silicate will foul the jets' steam nozzle and cause a loss of vacuum. This is the leading cause of loss of vacuum in most refinery vacuum distillation towers.

If the distillation tower's reflux, or bottom pumps, are steam turbine-driven pumps, the silicates will accumulate on the turbine blades and eventually break off. This will reduce the turbine's efficiency. But of greater consequence, this will also cause the turbine's rotor to become unbalanced and then result in severe turbine vibration and pump outage. This is the leading cause of shutdown of refinery turbine-driven pumps in my experience.

Reboiler Steam Condensate Drainage

1. The leading cause of poor distillation fractionation efficiency is certainly low reflux rates.
2. The leading cause of low reflux rates is certainly low reboiler duties.
3. The leading cause of low reboiler duties is low steam flow to the reboiler.
4. The leading cause of low steam flow to the reboiler is steam condensate back up in the reboiler Channel Head.
5. The leading cause of steam condensate back up in the Channel Head is that the condensate drain line is too small and too long.
6. The leading cause of the condensate drain line being too small and too long is that the condensate flashes through the drain line.
7. The flashing condensate increases the pressure drop in the drain line, which then causes more condensate to flash into steam.

Of course, any operator or unit engineer reading the above knows the industry-wide solution to this all too common problem. Which is to drain the condensate to the ditch. This is a wasteful practice, which increases the cost (about 20%) and the energy to generate steam, by not returning the condensate to the steam generation boiler.

Summary

The carry-over of boiler feed water into the steam supply to a distillation tower's reboiler is not particularly harmful as regards the extra moisture in the steam. The detrimental effects of water in the steam supply to a reboiler are the hardness deposits in the steam (i.e., the total dissolved solids or the TDS). This is related to the difficult problem of "Level Control" internally in the boiler itself, as compared to the external level indication, plus the accumulation of the hardness deposits at the steam-water interface in the boiler. These hardness deposits are concentrated at the top of the water level in the boiler, mainly in the form of silicates.

26

WATER-COOLED CONDENSERS

A horizontal, water-cooled distillation tower condenser is shown in Figure 26.1. Vapor enters the top of the shell, contacts the water-cooled tubes, and then condenses. The condensate flows out of the bottom of the shell. Cooling water flows into the bottom of the channel head, through the "U" shaped tubes, and exits from the top of the channel head. Sounds simple, but a variety of things often interfere with efficient condensation.

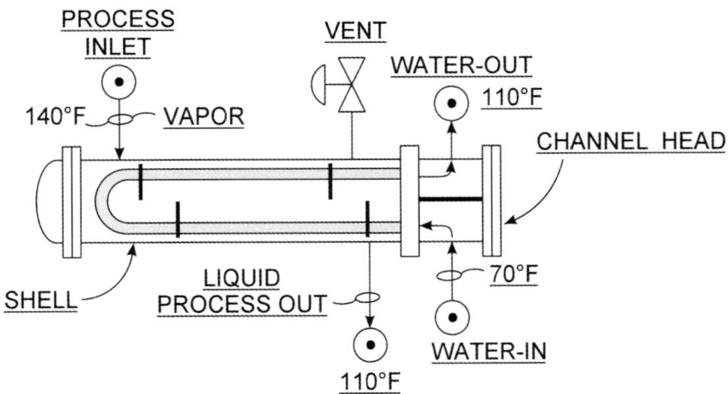

Figure 26.1. Water condenser

Condensate Backup

The most common problem with distillation tower condensers is liquid hydrocarbon condensate backing up inside the condenser shell. The tubes that are submerged in the condensate (i.e., liquid) cannot contact the condensing vapors. With fewer tubes exposed to the flowing vapor, the rate of vapor condensation slows. Any restriction to the condensate flowing out of the shell will result in condensate backup. Any increase in the elevation of the condensate drain line above the condenser will have the same effect. Even a wide-open control valve in the condensate liquid outlet may have the same detrimental effect that results in condensate backup. The most common problem I have experienced that causes condensate backup, is that the condensate drain line to the condensate collection drum is too long.

Non-Condensables

Saturated vapors such as propane, steam, naphtha, or butane often contain non-condensables such as:

- CO_2 in steam,
- Natural gas in butane,
- Air in naphtha (i.e., vacuum system condensers),
- Hydrogen in diesel desulfurizer effluent.

These non-condensables can become trapped at the top of the condenser's shell and reduce the number of tubes that come into contact with the condensable vapor. This is a common problem that can be corrected by opening the vent valve, which is shown in Figure 26.1.

The most common example of non-condensables reducing the rate of condensation is CO_2 contamination in steam. The boiler feed water flowing into the steam generator, or boiler, contains carbonates. Upon heating, the carbonates decompose into CO_2, which is a non-condensable. This CO_2 needs to be vented in steam reboiled distillation towers, by opening the vent beneath (not above) the channel head pass partition baffle. Opening the vent on the top of the channel head just vents steam and zero non-condensables.

Shell-Side Fouling

Often, the exterior of the condenser tubes is found to be fouled. In a refinery, these deposits are most often a consequence of air (i.e., oxygen) and olefinic hydrocarbons. The air is introduced into the condenser by exposing a portion of the feed in tankage to the atmosphere (i.e., when the tank lacks an inert gas blanket, such as nitrogen). The olefins are produced due to the thermal degradation of a portion of the condenser feed in an upstream process unit, at high temperatures.

When the oxygen (extracted from air) and the olefins combine, they form gums that adhere to, and foul, the exterior surface of the tubes. The best method to retard this type of exterior tube fouling is to avoid exposing the hydrocarbon feed to air in tankage. Blanketing a hydrocarbon feed tank with nitrogen is an effective method.

Tube-Side Fouling

A more common problem is that of tube-side fouling from the cooling water. Most often, the cooling water is circulated through a cooling tower. Due to hydrocarbon leaks in the water cooler's tubes, plus the air used in the cooling tower, sludge is formed in the circulating cooling water. This sludge then fouls the inside of the condenser tubes with what appears to be mud, but is really an organic sludge.

To get this sludge out of the inside of the tubes, we BACK FLUSH the tube bundle, typically, once a week, for about 20 minutes. The correct method to back flush a water cooler is:

- Step 1 – Open the water-side back-flush connection. This is a large valve that drains to the ground from the cooling water supply line.
- Step 2 – Close the water supply valve. The cooling water will now be flowing backwards through the tubes, from the cooling water return piping, and then onto the ground or sewer adjacent to the condenser.
- Step 3 – Wait until the water that is flowing into the sewer clears up from a muddy brown color to clear – typically, 10 to 20 minutes. Meanwhile, attach a plant-air hose (1") to the ¾" or 1" connection to the cooling water return line.
- Step 4 – With the water flow continuing, turn on the plant air. Immediately, the water flowing to the sewer will become muddy again. This air "Rumbling" effect agitates the dirt sticking inside the tubes and causes the deposits to break off the tube I.D.
- Step 5 – Wait until the water flowing into the sewer clears up again (another 5 to 10 minutes) and then return to normal operations.

The really effective way to keep condenser tubes clean on the water side is to find and fix the hydrocarbon tube leaks in the condensers. Checking the cooling water return line for hydrocarbons is the normal method to identify a tube leak.

Hardness deposits due to poor cooling tower water make-up chemical treatment is also a very common problem. These white solids are difficult to remove.

Chlorine Shocking

This is <u>not</u> the best way to improve condenser capacity. But it is the quickest, cheapest, short-term method to substantially increase condenser capacity that has been lost due to the cooling-water-side fouling of condenser tubes, due to sludge formation from the hydrocarbon leaks and/or calcium hardness deposits.

- Chlorine Shock Treatment – Establish a chlorine concentration (I use cylinders of chlorine gas) of <u>4 ppm</u> in the circulating water. Maintain this concentration for a period of 24 hours by adding chlorine.

- As the chlorine reacts with the hydrocarbon sludge, the cooling tower will foam and froth—which looks bad—but is just an indication that the chlorine shock treatment is effective.

Unfortunately, there is a long-term downside to this procedure. If your cooling water tower is constructed of wood (as my 12-cell cooling tower was in Texaco City), the chlorine will, over time, extract the lignite from the wood and over a period of several years cause the cooling tower to deteriorate.

Hot Vapor Bypass Control

Many of our distillation towers have their tower pressure control using "Hot Vapor Bypass." The tower pressure control valve is in parallel with the condenser and will open to increase the tower pressure. The vapor that bypasses the condenser flows directly into the reflux drum, where it increases the reflux drum pressure and temperature, and hence the upstream distillation tower pressure.

Perhaps the leading cause of the lack of condenser capacity in distillation service is a leaking "Hot Vapor Bypass." control valve. You can block-in the hot vapor bypass control valve and observe if the distillation tower pressure declines and if the reflux drum temperature also drops, to see if the bypass valve is leaking.

The usual cause of the leaking hot vapor bypass valve is an "O"-ring inside the valve (made of a pliable rubber) that becomes brittle with age and cracks. In 1970, I had such an old "O"-ring replaced for a cost of $70 and reduced the flaring losses from a depropanizer by $2,000/day at a refinery in Whiting, Indiana.

Tube-Side Water Velocity

Low water-side velocity will also promote tube-side fouling. Any tube-side water velocity of less than 3 ft./sec. is going to promote fouling. Doubling the tube-side velocity to 6 ft./sec. will greatly reduce, or even eliminate, much of the tube-side water fouling. But the water-side pressure drop will increase by a factor of "4." That is, the pressure drop varies with the velocity "squared."

Approach to "Wet Bulb" Temperature

To gauge how well your cooling tower is doing, we measure the "Approach to Wet Bulb." First call up your local weather station (or check your cell phone) for the current "Wet Bulb Temperature" which is a function of the relative humidity and ambient temperature. Let's say that it is <u>85°F.</u>

Compare this to your cooling water <u>outlet</u> temperature of <u>92°F</u>. You have a (92 – 85) = 7°F approach to wet bulb. The table below will determine how well your cooling tower is doing:

Table 26.1. – Approach to wet bulb validation

4°F	Excellent
6°F	Good—but can be improved (typical design)
8°F	Typical—but not good
10°F	Bad—but not unusual
14°F	Terrible

If the approach to wet bulb is bad, try cleaning the holes in the cooling tower's distribution decks and increase the air fan blade pitch to a maximum of a 23 degree angle. Also, make sure the fan belts are not slipping.

Recent Experience with the Overheating of Cooling Water

I have recently (July 2025) become an expert on the detrimental effects of extreme overheating of the cooling water at a refinery close to my home in Louisiana. On a crude unit distillation tower, the cooling water outlet reached 165°F. The result was extreme fouling of the condenser, cooling water return line, and especially of the cooling tower itself. What appeared to be whitish calcium deposits had fouled the cooling water return line to an extent that the return water flow was restricted. Of greater significance were the horrendous hardness deposits in the cooling tower itself. These deposits had distorted the return water distribution to such an extent that the cooling tower performance was so compromised that the "Approach Temperature to Wet Bulb" of the cooled water was 20°F, vs. the design of 6°F approach to wet bulb. This problem was the main bottleneck in the performance of the crude distillation tower.

Summary

Fouling of water-cooled condensers is most often caused by overheating the cooling water return temperature above 125°F. Excessive cooling water return temperatures above 150°F are catastrophic.

A cooling tower approach to wet bulb temperature of 6°F to 8°F is okay. A cooling tower approach to wet bulb above 10°F is terrible.

Chlorine shocking the circulating cooling water is an effective method to remove organic fouling deposits from cooling water exchangers, but also causes cooling tower deterioration, if the cooling tower is constructed of wood, over a period of years.

Hardness deposits in the cooling water return line are a sure sign of excessive cooling water return temperature.

27

FIN-FAN AIR CONDENSERS

All air coolers use finned tubes, as shown in Figure 27.1. The tube O.D. is typically 1" to 1½". The tubes are densely covered by 1"-wide, very thin fins. The fins increase the outside surface area of the tubes by a factor of about 12. These fins are very easily bent, as they are made of thin aluminum. The fins do not increase the cooler heat transfer heat rate by a factor of 12, but by a factor of 3 or 4. Or sometimes, by much less.

In most air coolers, the fan is located below the finned tube bundle. That's called a forced-draft fan. Perhaps in 10% to 20% of air coolers, the fan is located above the tube bundle. That's called an induced-draft fan. The reason for the fans to be mainly located below the bundle is for ease of maintenance. Process considerations alone would lead us to use induced-draft fans.

Tube-Side Passes

The majority of tube bundles are the two-pass configuration shown in Figure 27.1. The fluid to be cooled or condensed enters at the top of the channel head and then flows through the top half of the tubes, turns around at the end of the bundle, in the return header box, and then flows back to the channel head, through the lower half of the tube bundle.

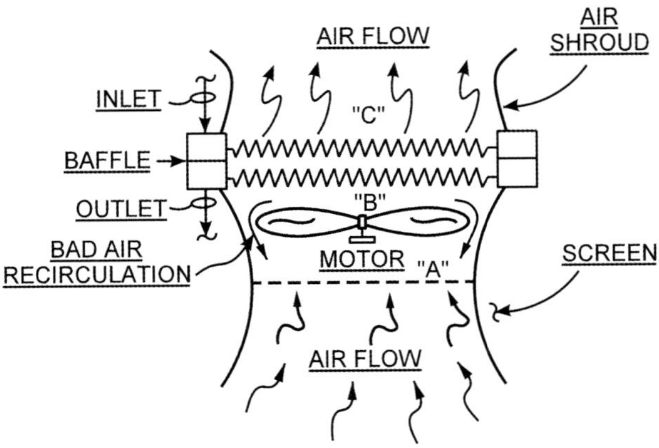

Figure 27.1. Two-pass configuration

The fan is normally spun by a motor. Typically, the motor is a fixed speed, 3-phase A.C. motor, rotating at 3,600 rpm (North America) and 3,000 rpm (Europe, Asia, Africa). The fan spins much slower than the motor. The motor and fan are connected by a drive belt.

Increasing Air Flow

The simplest method to increase the air cooler capacity and distillation tower condenser capacity is to increase the air flow by making the fan run faster. Placing a bigger wheel on the shaft of the motor will make the fan spin faster. But this also increases the amp load on the motor driver as follows:

- Increasing the wheel size by 10% increases motor amps by 34%.
- Increasing the wheel size by 20% increases motor amps by 73% (i.e., work varies with speed to the 3rd power).

So, before placing a bigger wheel on the motor shaft, check that the motor's "Full Limit Amp" load will not be exceeded. Check the current amp load on each of the three wires connected to the motor.

A simpler method to get more air flow is to increase the angle of the blade pitch. The maximum angle is 23 degrees. A minimum angle is about 8 to 10 degrees. It only takes a few moments to increase the blade angle by loosening two bolts. But, going from 10 to 23 degrees only increases air flow perhaps by 20% to 30%, at most.

One annoying problem with increasing the blade pitch to increase the air flow is that reduced ambient temperatures require more motor amps. Thus, if we increase the air flow with a bigger wheel on the motor shaft, or by setting the blades at a steeper angle, the motor may trip-off on excessive amps, if it gets colder at night and the cold air density increases. This is a big problem is desert areas, where night time temperatures may drop a lot from daytime peaks. I learned this from personal experience, when I worked in the desert region of Laredo, Texas, in the mid-1980s, where the ambient temperature might decline from 105°F during the day to 50°F at night.

Fin Tube Fouling

Referring to Figure 27.1, the air flows at points "A" and "B" are the same. But, as the tube's external fins foul with dead moths and willow seeds, the air flow at point "C" is reduced. The difference in the air flow between points "B" and "C" is shown on Figure 27.1 as "Air Recirculation." Therefore, as the fins foul, the amp load on the motor driver remains about the same, based on my experience.

One indication of the loss of air flow is an increase in the temperature rise between points "B" and "C" (see Figure 27.1).

Restoring Air Flow

1. The simplest (but not necessarily the most effective) method of restoring air flow lost due to external tube fin fouling is to shut off the fan, remove the screen (see Figure 27.1), and sweep the bottom two rows of finned tubes with a long handle broom.
2. A more effective method (assuming the motor is a 3-phase and not a 2-phase motor) is to reverse the polarity of the electric leads. This will cause the fan to run backwards and air to be blown backwards through the fins. This will blow the dirt and dust off of the bottom two rows of finned tubes, but perhaps into the intake of any adjacent air coolers?
3. My favorite method is to use a jet of water from a hose with 20 to 30 psi of pressure and to squirt the water between the bottom two rows of fins. I prefer to do these sorts of things myself, so as to avoid bending the fins, which are very easily deformed if not treated very gently! The fins are made out of very thin aluminum and are typically one inch wide. However, standing back too far, or using too little water pressure, will prevent washing the willow seeds, road dust, and dead moths from the area between the lower rows of fins and the exterior tube wall.

Water Mist

An effective method to reduce an air cooler process outlet temperature by 4°F to 8°F (but only in climates with a low relative humidity) is to introduce a water mist beneath the fans. This water flows through a mist nozzle located a few feet BELOW the forced draft fan, so that the water entirely evaporates before it impacts the tubes. Typically, this reduces the air outlet temperatures from the fans by 10°F, but only if the air is dry (i.e., a low relative humidity, as in south Texas).

Operators should turn on the water mist midday, as ambient temperatures rise, and shut off the water at sunset. Use a 1” nozzle or larger, and only use clean water, to keep the nozzle from plugging. I use 2” nozzles to be on the safe side.

I learned this trick in Lithuania. It was standard refinery technology in the defunct Soviet Union.

Tube-Side Pass Configuration

Note that the cooler shown in Figure 27.1 is a 2-pass. That is, the process inlet and outlets are on the same end of the cooler. The process fluid reverses its flow in the return header box, at the far end of the exchanger. This type of arrangement is most common because it permits the back end of the exchanger to expand or contract as the tubes heat or cool. For this reason, the return header box is not rigidly attached to the cooler support structure, but is allowed to expand and contract freely.

Unfortunately, leakage of the baffle shown in Figure 27.1 (a common problem) will then reduce heat transfer efficiency by a very large amount. Inspection of the mechanical integrity of this baffle is often very difficult. In most plants in North America that I have worked in, the inlet channel header box is not removable. However, in Europe, they typically are. Why, I don't know.

A True Story from Mumbai, India

I was leading our Troubleshooting Seminar in Mumbai, India, about 15 years ago, when one of the attendees said, "Mr. Norm, we have a forced draft air cooler that has a low cooling efficiency. Could you look at it with us after class today?"

That evening, the entire class went to look at this air cooler, which was cooling vacuum tower bottoms. That means asphalt, used to pave roads. The process inlet temperature was 450°F, and the process outlet temperature was 448°F. The return header box (Figure 27.1) was cold to my touch. And the pressure drop of the flowing asphalt was zero! The baffle (shown in Figure 27.1) had been ruptured, because the asphalt had solidified inside the tubes, one minute after it had been put into service, ten years ago. A true story.

Summary

Most air coolers have the fan located beneath the tube bundle for ease of maintenance. The best way to increase air flow is to squirt water carefully between the bottom two rows of tubes, between the 1" aluminum fins, being careful not to bend these fins.

Spraying water into the air flow, in climates with a low relative humidity, is another effective method to reduce the process outlet temperature by 3°F to 4°F.

Making a fan run faster by increasing the wheel size on the motor shaft is a very effective method to increase air cooler capacity, provided that the fan's motor driver is not overloaded. The amp load increases with the wheel size to the 3rd power. Compare these to the full limit amp load on the motor driver.

Before increasing the fan speed, check with the company that provided the air cooler to make certain that the maximum allowable torque load on the fan blades has not been exceeded. This is, I believe, a legal safety requirement – at least in Texas.

28

PUMPAROUND HEAT EXCHANGERS

The main purpose of the pumparound (Figure 28.1) is to preheat the crude charge with the circulating hotter diesel from tray 29. The heat recovered in the P/A reduces the heat load of the overhead condenser of the fractionator and also reduces the firing rate of the crude preheat furnace, which saves energy.

But do the four trays that I show in Figure 28.1 also contribute to fractionation efficiency? The answer is yes. But by how much? This is a complex issue that is not well understood in our process industry.

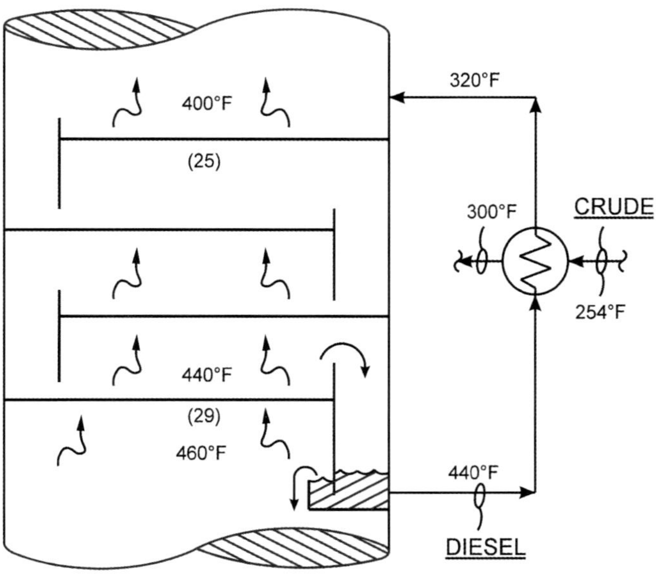

Figure 28.1. Pumparound

Effect of Pumparound on Fractionation Efficiency

Increasing the pumparound duty has two potential benefits:

1. It reduces the vapor flow to the trays above the pumparound return tray, and may stop tray flooding on trays above the pumparound return.
2. It recovers heat that can be used to generate steam in a waste heat steam boiler or to increase the crude preheat temperature.

However, increasing the pumparound duty has two potential debits:

1. It reduces the tower top reflux rate and may therefore reduce fractionation efficiency on the trays above the pumparound section.
2. The increased pumparound circulation rate may cause the pumparound section trays to flood, due to excessive downcomer liquid loadings, or because of excessive tray weir loadings.

Crude distillation towers, vacuum towers, fluid cat crackers, and delayed coker fractionators all have several pumparounds. Typically, each pumparound is considered to represent a single, 100% efficient, theoretical fractionation stage, or one 100% efficient fractionation tray.

This industry-wide assumption is usually wrong! Let's say that our diesel oil pumparound has four trays. Assume that we observe (Figure 28.1) that the:

1. Temperature of the vapor leaving the top tray is 400°F.
2. Temperature of the vapor leaving the bottom tray is 440°F. Therefore, the four trays represent more than one theoretical fractionation stage, as indicated by the temperature drop of the vapor, as compared to the temperature increase of the liquid that leaves the pumparound section.
   - Vapor leaving the top P/A tray is 400°F
   - Liquid leaving bottom P/A tray is 440°F

Then, these four trays represent more than a single theoretical fractionation stage, or more fractionation efficiency than a single tray with 100% efficiency.

Computer Simulation vs. Field Observations

You can see this effect in a tray-to-tray computer simulation, if the simulation assumes multiple theoretical stages for the pumparound sections. Also, you can observe this effect from a fractionator field temperature survey, using an infrared surface temperature gun, as I have done many times myself.

While it's true that trays in pumparound service don't contribute as much to fractionation efficiency as do the other trays, they do have a significant effect on fractionation between adjacent side draw-off products.

Pumparound Heat Removal

The principal purpose of the pumparound shown in Figure 28.1 is to remove heat from an intermediate section of a fractionator. Hot diesel is drawn from tray 29, cooled by heat exchange with crude oil, and then returned to tray 25.

The effect of increasing a pumparound circulation rate may increase the heat extracted from the tower. Or it may decrease the heat extracted from the tower. This is a complex problem, and I have explained this important problem in detail below.

Effect of Increasing Pumparound Flow Rate

1. Increase P/A flow.
2. P/A draw temperature will always decrease.
3. P/A return temperature will always increase.
4. The temperature difference between the draw temperature and return temperature will always decrease.

If the decrease in the P/A draw-off temperature, minus the P/A return temperature, has decreased by a greater percent than the increase in the P/A circulation rate, then you will be removing "less" heat with the increase in P/A flow. I see this happening on fractionators perhaps 20% to 30% of the time.

Panel Board Operator Error

Panel board operators make this mistake frequently. They will increase the P/A circulation rate to remove more heat from their fractionator, which it normally does. But very often, this increase in P/A circulation rate reduces the heat extracted from the fractionator and therefore increases the vapor flow rising from the P/A circuit, which may cause flooding on the trays above the pumparound return.

Pumparound Exchanger Problems

1. A floating head allows for expansion of tubes, as the tubes heat up.
2. But a floating head does not allow for the different rates of thermal expansion of individual tubes, which may get plugged.
3. When a tube plugs with crude on the tube side, those tubes with low flow must expand more than the colder tubes that have not plugged.
4. But the cold tubes will prevent the hotter tubes from expanding by preventing the movement of the floating head.

5. Therefore, the hotter tubes are distorted due to the differential rate of thermal expansion between the hot tubes and the cold tubes.
6. This distorts the shell-side flow and causes reduction in heat transfer efficiency due to dead spots on the exchanger's shell side.
7. To minimize this problem, we need to put higher-viscosity, dirtier material (i.e., crude) on the exchanger's shell side, as shown in Figure 28.2.

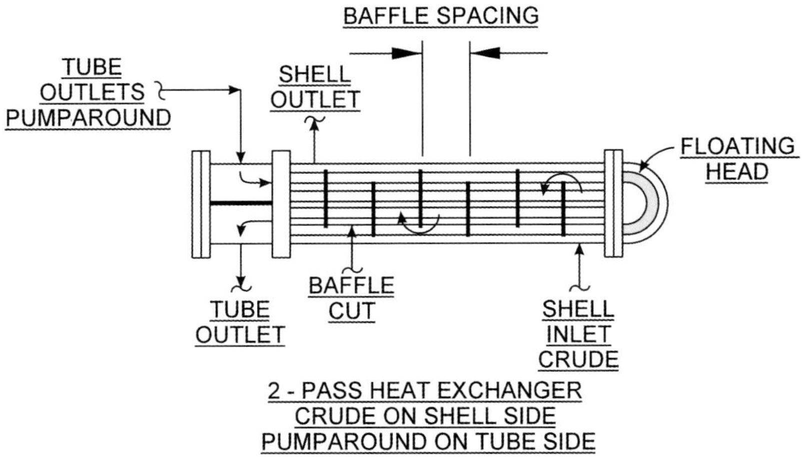

Figure 28.2. Pass heat exchanger crude on the shell side pumparound on the tube side

Industry Standard Configuration

The industry standard is to place crude oil on the tube side and the clean products, or P/A, on the shell side. The reason for this is that it's difficult to clean the shell side due to:

- Triangular tube pitch
- ¾" tubes
- 1" tube spacing

Therefore, there are no cleaning channels on the shell side.

My standard is to place the crude oil on the shell side and the cleaner, less viscous products on the tube side. But then, to permit cleaning of the shell side of the heat exchangers during a turnaround, I use the following tube configuration:

- Rotated square pitch
- 1" tubes
- 1½" tube spacing

This conforms to a very old Exxon (then Esso) standard from the 1930s to 1940s that I observed at a former Esso refinery in Aruba designed prior to World War II. Basically, the idea is to place the higher-viscosity material on the shell side of an exchanger, so as to take advantage of vortex shedding, as a liquid flows at right angles across a tube. Vortex shedding greatly reduces heat transfer film resistance. This old Esso standard is correct! I adhere to this crude exchanger configuration in all my crude unit retrofit designs. It really works great!

Summary

My story about the effects of pumparound circulation rates on the tower's heat balance and fractionation efficiency illustrates a complex issue that is never taught in universities. It's an example of the importance of learning process technology from personal and direct field observations. The key component in becoming a competent process operator or an effective plant engineer is direct and personal field observations and an analysis of the observed data. University education is of secondary importance to the complex job of being a competent practicing chemical engineer. I have discussed the complexity of the design and operation of pumparounds in the following chapters, relating to optimizing pumparounds to save energy and how to correctly represent a P/A on a computer tower distillation program. A difficult problem!

29

OPTIMIZING PUMPAROUNDS TO SAVE ENERGY

The concepts I'll describe in this chapter will work on multi-component fractionators, including crude units. It will also increase the heat recovery for:

- FCU fractionators
- Visbreaker towers
- Hydrocracker fractionators
- Vacuum towers
- Delayed coker fractionators

These new concepts will GREATLY improve the energy efficiency compared to standard industry practices for pumparound heat recoveries.

Stripper Feed Side Draw-Off Location

Let's assume we are making a heavy gas oil side draw-off product flow to the side-stream steam stripper from a total trap-out chimney tray (Figure 29.1). The feed to the stripper consists of two parts:

1. The gas oil that condenses in the P/A, which I will call stream "A" (Figure 29.1).
2. The gas oil that flows onto the top P/A tray from the tray above (which I'll call stream "B") – that is, the internal reflux flowing down the tower from the packed section above.

Allowing stream "B" to mix inside the fractionator with stream "A" reduces the average boiling point of the circulating P/A. This also reduces the P/A draw-off temperature. Of course, this is what normally takes place in a distillation tower. However, my design, as explained in Figure 29.1 and Figure 29.2, avoids this loss of fractionation efficiency and also the degradation of the pumparound draw temperature.

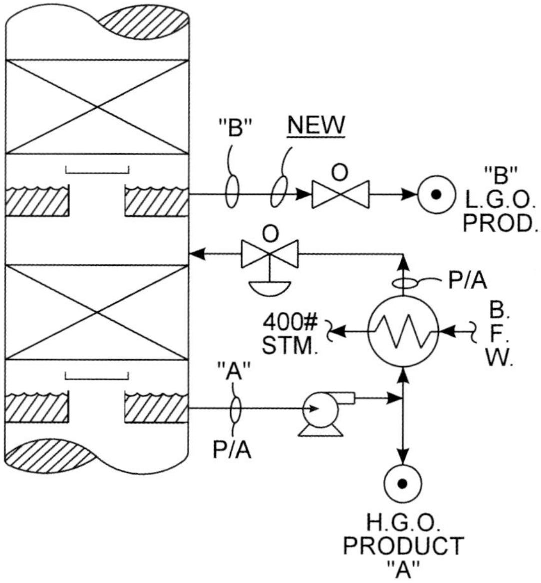

Figure 29.1. LGO draw from Nozzle B increases Nozzle A pumparound temperature and improves heat transfer efficiency

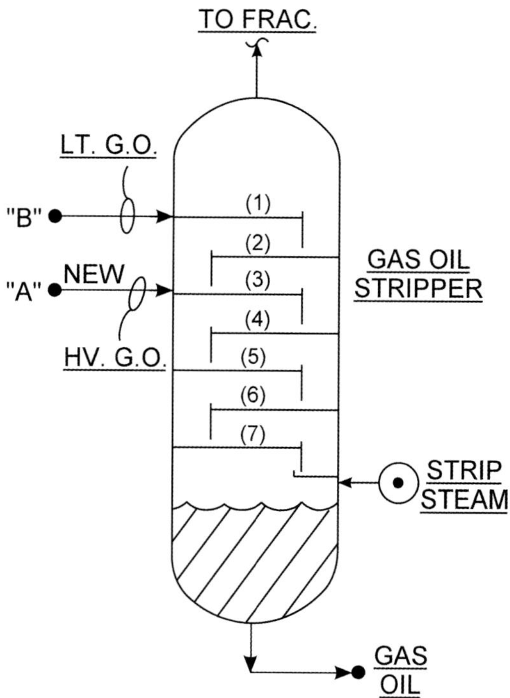

Figure 29.2. Introduce LGO two trays above HGO to improve stripping efficiency and diesel recovery

To raise the P/A draw-off temperature, I will TOTALLY draw off the internal reflux (Figure 29.1 - stream "B") above the P/A return tray. I will then feed it into the side-stream stripper, two trays ABOVE the tray, where stream "A" was introduced into the steam stripper, as shown in Figure 29.2. This complexity will save lots of energy and is worth your extra effort to understand. I first used this novel technology at a very large refinery in Texas City, on a new delayed coker fractionator that I had designed in 1966. This was the first fractionator that was ever designed with multiple pumparounds, using a computer simulation. It is still in operation.

Effect on Fractionation

The introduction of the lighter stream "B" two trays above where the heavier stream "A" was introduced had, on my computer simulation, a positive effect on the steam stripper fractionation efficiency, as I will explain in the following section. I had to create my own simulation program in 1966.

Pumparound Fractionation Efficiency

There is a widespread misconception that a pumparound represents only a single theoretical fractionation stage. This, in general, is wrong, because of the following three factors:

1. If the temperature of the vapor leaving the top of a pumparound is the SAME as the temperature of liquid leaving the bottom of the pumparound, then the pumparound does indeed represent a single theoretical fractionation stage.
2. If the temperature of the vapor leaving the top of the pumparound is cooler than the temperature of the liquid leaving the bottom of the pumparound, then the pumparound is developing MORE than one theoretical fractionation stage.
3. If the temperature of the vapor leaving the top of the pumparound is hotter than the temperature of the liquid leaving the bottom of the pumparound, then the pumparound is developing LESS than one theoretical fractionation stage.

On a computer simulation, I typically represent a pumparound with 2 or 3 stages and use 3 to 5 actual trays in the fractionator I am designing. To calculate a fractionator's pumparound heat extraction, we have to multiply two factors together:

1. The increase in the pumparound circulation rate ("Delta F"),
2. The pumparound draw-off temperature, minus the pumparound return temperature (Delta "T").

(Delta F) • (Delta T) = Delta Q

Where "Delta Q" is the change in the pumparound duty.

When we increase the pumparound circulation rate, the pumparound draw-off temperature will always go down. But the pumparound return temperature will always go up, unless we were to add more heat exchangers to the pumparound circulation loop. Usually, increasing the pumparound flow will increase the pumparound heat duty, but not always! Sometimes, both in theory and as can be observed on an operating fractionator, the pumparound duty will go down as I increase circulation rates!

It's a complex situation, like everything else in process operations.

I have since used this concept in crude and FCU fractionators, as well as the four other delayed coker fractionators that I have designed since 1966, to maximize the pumparound draw-off temperatures, and thus to improve energy efficiency in complex distillation towers.

Summary

The dual draw-off concept I have detailed above will improve fractionation and will also improve heat recovery from pumparounds. However, this concept has not been adopted by the refining industry, in general, even though I described it in my first book, published by PennWell Books in 1980, *Troubleshooting Process Operations*, which is currently available in its 5th edition.

I invite the reader to perform a computer simulation of this concept, and to draw their own conclusions. It's a way to save energy and to also enhance distillation tower fractionation efficiency. I have used this concept at a refinery in Sugar Creek, Missouri, to increase the pumparound draw-off temperature about 50°F (Figure 29.1), with excellent results. This true story also is an example of the benefits of computer simulations for new Distillation Tower designs, as discussed in the following chapter.

30

COMPUTER SIMULATION TO DESIGN A DISTILLATION TOWER

Early in my career, I designed one of the very first fractionators with multiple side draw-offs in the refining industry's history, using a tray-to-tray computer simulation. That was in 1966, in the days of "punch cards," when I had to write the tray-to-tray computer program myself, long before off-the-shelf programs were available.

It was the "B" coker fractionator, for the new delayed coker, for a refinery in Texas City. Each tray was represented by a single flash drum. In my computer simulation, that represented a fractionation stage of 100% efficiency.

There were no conversion programs. I just used a "Do Loop" that recycled the feed, reflux, and stripping steam 100 times. I would then examine the output, readjust the inputs, and let the program run for another hundred "Do Loops." Each run required about 30 minutes of computer time in 1966. There were no converged solutions. It was just a matter of my re-running my "Do Loop" program and then readjusting my starting point, once a day, for a few weeks.

Eventually, after several weeks, I produced a vessel sketch for the new "B" coker fractionator (Figure 30.1). Nine months later, the tower was constructed and put on-line. It worked quite close to my computer-generated design. These were the products:

- Off-gas
- Naphtha (light)
- Naphtha (heavy)
- Light gas oil (diesel)
- Heavy gas oil (FCU feed)
- Recycling of heavy gas oil to the coke drums
- Delayed coke product

Lab results were right on spec and very close to my projected flow rates and my design draw-off temperatures and composition. This was quite a success for me, for the very first delayed coker fractionator designed with a computer in the history of the refining industry!

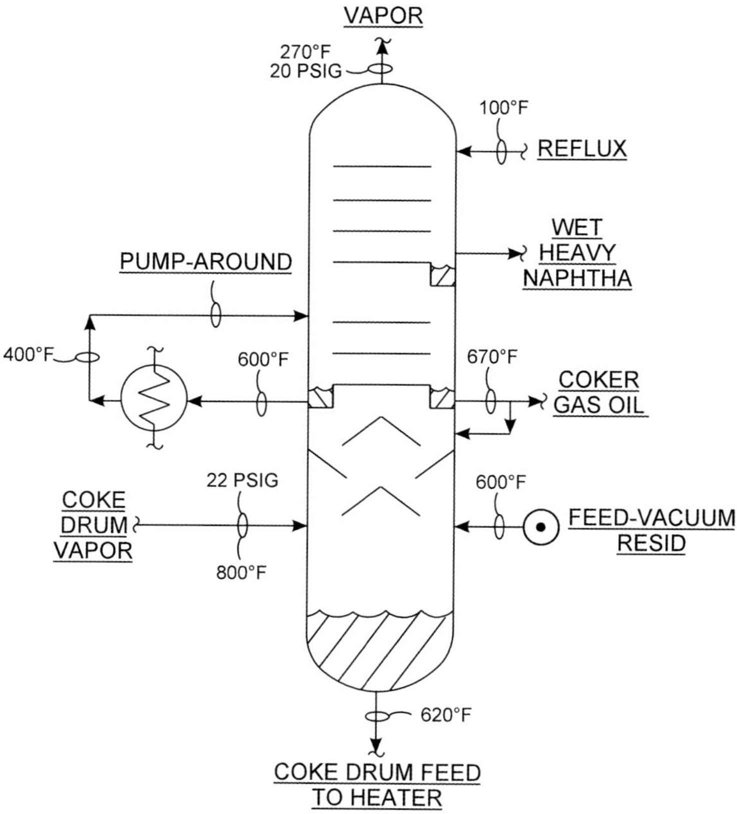

Figure 30.1. Coker fractionator

What was the secret of the first successful design using a computer simulation?

The old "A" coker was already in operation in the refinery in Texas City. Before I wrote my new computer program for the "B" delayed coker fractionator, I wrote a tray-to-tray computer simulation for the old "A" coker. I then adjusted all the parameters of my new computer simulation, based on the plant test data I had observed for the operation of the old "A" delayed coker fractionator, which had been in service for over ten years.

A computer simulation needs to be based on a real-world prototype to be really useful.

My Current Practice

If you worked with me today and observed how I go about designing a new distillation tower to fractionate crude, LPG, and reactor effluents, you would observe that I still rely on that technique used in 1966. That being, I select the relevant "Equation of State" for the computer model of my current design, based on a prototype of a distillation tower in similar service.

If you do not have a firm basis to select the equation of state (which is used to calculate the relative volatility of the key components), then you are not really designing your new distillation tower. What you are actually doing is guessing at the number of trays, at the reflux rate, the reboiler duty, and the tower diameter, and all of the other distillation parameters. After almost 60 years of refinery distillation tower design experience, I have found that the incorrect selection of the equation of state used for the tower design, is one of the main factors in poor performance of new distillation towers.

Naphtha Side Draw-Off Product Water Contamination

To conclude this description of my first computer model-based design, I personally participated in the start-up of the new "B" delayed coker unit in Texas City in 1967. The new "B" coker had a feature that the old "A" coker lacked (Figure 30.1). This was a Heavy Coker Naphtha side draw-off product. During the coker start-up, I drew a sample of the heavy naphtha in a glass bottle. Much to my surprise, my sample was about 10% water.

I then reviewed my design computer simulation result from the prior year. And to my surprise, the simulation output also showed that the heavy naphtha side draw-off product was contaminated with several percent of free water. It was something that I had missed when applying my computer simulation output to my detailed design of the fractionator. That is, the need to keep the tower top temperature hot enough to prevent condensation of any water inside the tower, to retard corrosion.

The industry rule of thumb is to keep the tower top temperature at least 30°F above the calculated water dew-point temperature. The condensation of any water promotes fouling and corrosion of the top few trays of the distillation tower in FCU, crude oil, and coker fractionators. For this reason, the only delayed coker fractionator ever built in the world with a heavy naphtha side draw-off was the "B" coker in Texas City in 1966. A very bad mistake that I have not repeated. There is just too much water vapor coming from the upstream coke drums to the coker fractionator to avoid water condensation inside the tower. Unless the tower's top temperature is kept hot enough to take all of the coker naphtha overhead as a product, as was correctly indicated by my primitive computer simulation for this distillation tower back in 1966. But, I'm smarter now!

Summary

Using a computer simulation to design distillation towers is now standard industry practice and has been for the last half century. But the selection of the proper equation of state, which is used to determine the relative volatility of the key components, remains an engineering challenge. This problem is too often ignored by design engineers, who just use the default value in their computer simulation. This is not the correct approach to a complex problem. Always base your design simulation on a performance test run on an existing distillation tower, in a similar service.

Safety Note – Unplugging Bleeders

We have all seen the unsafe way to unplug a bleeder, in hydrocarbon services. Occasionally, this results in a bleeder "blowing out" and emitting a flammable, toxic vapor. The safe way to unplug a bleeder is to have the maintenance personnel, or the operators, obtain a "Rod Out Tool." This is the standard way to safely plug bleeders in hydrocarbon services. This tool tells the operator when they successfully unplugged a bleeder and when it is safe to remove the rod-out tool from the unplugged bleeder.

Every plant I've worked in has a rod-out tool available. It only takes a minute to use it to unplug a bleeder SAFELY.

31

TROUBLESHOOTING USING RADIATION SCANS

Determining the exact tray, or the packed section location where flooding is initiated in a distillation tower, has become quite straightforward in the last few decades. This is done very effectively with a "Radiation Scan." The result of such a scan is shown in Figure 31.1. It is a simple and effective method to precisely determine where the flooding is being initiated in both trayed and packed distillation towers. It is a truly excellent distillation tower troubleshooting tool for both hydrocarbon and aqueous services.

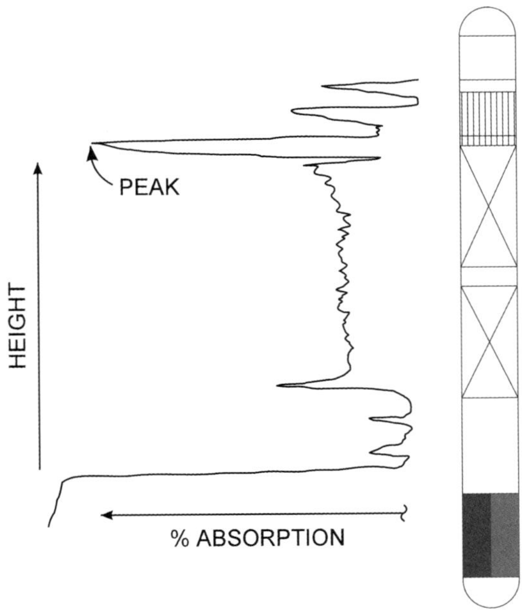

Figure 31.1. Radiation scan. Top of packed bed is causing flooding

How It Works

A source of radiation is placed on one side of a distillation column. The radiation receiver is placed on the opposite side of the tower. The radiation source and the radiation receiver are then moved in unison along the length of the tower. And, the percent of absorption of the radiation versus elevation is then recorded, as shown in Figure 31.1.

It's simple, reliable, and very easy to interpret the result. It determines with precision the exact elevation in the fractionator, where flooding is initiated.

Why I Do Not Initiate Use of Radiation Scans

If I am provided with a radiation scan, such as illustrated in Figure 31.1, I will gladly use it with confidence. However, I have never initiated use of this valuable troubleshooting tool myself. Why?

- It's expensive at perhaps $15,000 to $35,000, or much more. However, quite recently, my clients report that the cost of a typical scan has dropped to less than $15,000.
- From the time a request is made, until the results are reported by the company executing the scan, typically four weeks or even longer may pass. Even for me to have to wait four hours would be too long. Also, I can't do it myself, which for me, is a major problem.

Typical Scan Results

Figure 31.1 shows a typical scan result. The scan was taken on a refinery crude tower kerosene, side-stream steam stripper that was known to be flooding. That is, when stripping steam flow was increased, the tower bottom flow was greatly reduced, and the bottom level went to zero.

The scan definitively shows that the flooding is initiated at the top of the upper bed of packing, most likely as a consequence of corrosion, of the fouling deposit accumulation on the packed bed top few inches. This is an extremely common problem in crude distillation service, caused by:

$$HCl + Fe \rightarrow Fe\,(Cl)_2$$

$$Fe\,(Cl)_2 + H_2S \rightarrow Fe\,(HS)_2 + HCl$$

This reaction only can occur in an aqueous environment. The $Fe(HS)_2$ (which is called pyrophoric iron) is a black solid, and is the most common cause of fouling and flooding in the upper parts of crude distillation towers, and in many other refinery distillation services.

Radiation in Level Control

Radiation is often used in distillation towers for bottom level control. Typically, this is a more reliable method for level control in the bottom of distillation towers when:

- Fouling,
- Foaming,

of the bottom liquid level product is a problem, as compared to conventional level indications and control.

Care must be taken to handle the radioactive level indication sources, in accordance with rather precise federal legal restrictions. I inadvertently lost a radioactive level indicator on a spent sulfuric acid tank in Texas City on my alkylation unit in 1974, and I was almost discharged as a result by my employer. It was certainly not my fault. I directed the maintenance supervisor to have the tank cleaned of spent sulfuric acid sludge. And then, I forgot all about the work order I had signed. That was 52 years ago. I am much smarter now as regards blaming other people for my negligence.

On a more recent project, in an amine fractionator, I used a scan to clearly identify that most of the liquid flow was by passing around the periphery of a bed of structured packing. Most likely, this was because the structured packing was NOT in good contact with the vessel wall, which is a rather common shortcoming in the use of beds of structured packing in distillation services. A special, very thin, angled metal ring is used around the sections of the structured packing, to reduce liquid and vapor by passing, along the vessel wall inside diameter.

Summary

I am a tremendous fan of radiation scans as a distillation tower troubleshooting tool. Recently, the radiation scan technology has advanced in that the scan can now look down to obtain the TOP VIEW of a cross-section of the tower. This is a really wonderful new feature of scan technology.

Also, the cost of these scans in the past years has greatly diminished, to about half of the prior cost. There is absolutely no better tool to aid us in our understanding of the performance of a distillation tower than the tray-to-tray radiation scan.

The real problem with use of these scans is the time it takes to get the scan run, sometimes even after management approval. However, on balance, radiation scans are almost always worth the extra time and the extra money.

32

SAVING ENERGY IN DISTILLATION

When operating a distillation tower, I focus on ways to reduce the steam flow to the reboiler, with the objective of energy savings. The simplest thing to try is to increase the feed preheat, if the feed preheat comes from a source of waste heat, such as low-pressure steam that is vented to the atmosphere.

My usual approach is to optimize tower operating pressure. A reduction in operating pressure will increase the "Relative Volatility" of the key components. The definition of relative volatility (RV) is: vapor pressure of the light key component (VP_L) divided by vapor pressure of the heavy key component (VP_H).

$$RV = VP_L \div VP_H$$

Look at a vapor pressure chart curve for propane and butane. Note that the curves spread apart, like the fingers on my hand, as their vapor pressures and boiling point temperatures decrease.

Tower Temperature Profile

Unfortunately, increasing feed preheat with low-pressure steam, and a reduction in the tower's operating pressure, may reduce fractionation efficiency at a fixed reflux rate. Increasing feed preheat, at a fixed reflux rate, will certainly reduce the reboiler heat required. But this will also unfortunately reduce the stripping factor (S):

$$S = VK \div L$$

Where:

V = moles of vapor flow

L = moles of liquid flow

K = the vapor pressure of the tower overhead product, divided by the tower pressure

As we reduce the reboiler duty, this stripping factor is also reduced.

As we reduce the stripping factor, the tray effectiveness below the feed tray is diminished. That is, trays below the feed point are not working as hard to strip the lighter components out of the feed.

Entrainment

But the problem is more complex than what we learned in college, or in operator training class.

Why?

Because the trays above the feed point might be suffering from entrainment. That is, the vapor blowing through the tray decks had sufficient velocity to lift the liquid from a lower tray up to the tray above. This will compromise the efficiency of the upper trays, due to entrainment.

The simple solution to this loss of fractionation efficiency, would then be to reduce the reboiler duty, to reduce the entrainment on the trays above the tower's feed tray.

But if the reboiler duty is reduced too much, then the tray decks will start to leak, or weep, or dump. That is, the liquid will drain through the lower points of the tray deck, and the vapor will flow to the higher points of the tray. This will reduce tray fractionation efficiency, due to the loss of vapor-liquid contacting below the tower's feed point.

Using Temperature Profile to Save Energy

I had this problem one night in 1981 at a refinery in Louisiana. How could I determine, as I increased the 30 psig steam to the propane-butane splitter (Figure 32.1) at a fixed 100 psig steam flow to the reboiler, whether fractionation between the two products: propane (going to LPG product sales) and butane (going to gasoline blending for RVP control) were getting better or worse? (RVP = Reid Vapor Pressure)

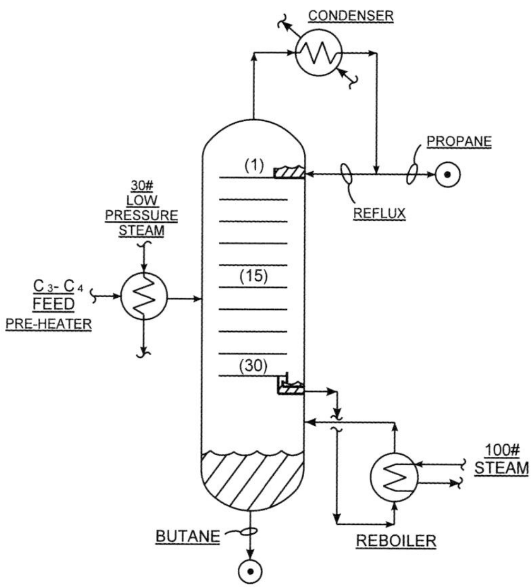

Figure 32.1. Propane-butane splitter

It was late, and the refinery lab was closed. Samples were not an option. So, I relied on my favorite tool: the TEMPERATURE PROFILE, that is, the temperature difference (DT) between the tower top vapor and the tower bottoms product. If this DT goes up, then fractionation is getting better. If the DT diminishes as I increase the feed preheat at a constant reflux rate and a reduced reboiler duty, then fractionation between the propane and butane is going to get worse.

So I had the panel board operator:

1. Increase feed preheat
2. Keep the reflux rate constant
3. Lower the steam flow to the reboiler, to keep the reflux rate constant

And the temperature difference between the top and bottom of the tower remained the SAME, which was good.

Why?

Because the trays in the lower half of the tower had been suffering from a loss of fractionation efficiency, due to entrainment, before I increased the feed preheat!

That night, I reduced the high-pressure steam flow to the reboiler by 10,000 L/H, by using 10,000 LB/HR of low-pressure steam to the feed preheating that had been vented to the atmosphere and wasted. So, I reduced refinery fuel usage at the boiler plant by about 12,000,000 BTU/HR, which is a significant energy savings.

As discussed above, lowering the tower pressure enhances the RELATIVE VOLATILITY of the key components and makes separations easier. This means less reflux and less reboiler steam flow is required. As you do this, and the temperature difference between the top and bottom of the tower increases, you're moving in the right direction.

But if this temperature difference diminishes, you're losing fractionation efficiency due to entrainment, or tray flooding. So, you have gone too far in your effort to save energy by reducing the tower's operating pressure.

Condenser Capacity Limits

Efforts to reduce the fractionator operating pressure to save energy by reducing reflux and reboiler duty are always limited by condenser capacity, as follows:

1. If you have hot vapor bypass pressure control and do have an excess of condenser capacity, the hot vapor bypass control valve will be partly OPEN. If it's closed, you cannot reduce tower pressure.
2. If you have flooded condenser pressure control and do have an excess of condenser capacity, then the reflux drum will be completely FULL, and you can then reduce the tower operating pressure until a level appears in the reflux drum. At this point, the condenser is completely un-flooded, and all of the tubes are exposed to the condensing vapors.

Saving Energy in Distillation Towers Checklist

1. Check tray deck levelness.
2. Check weir levelness.
3. Optimize tower operating pressure to maximize top-to-bottom temperature difference, at a fixed reflux rate.
4. Only use once-thru thermosyphon reboilers. Never circulating thermosyphon reboilers, where the bottoms product is recirculated through the reboiler.
5. Keep the reboiler start-up valve completely CLOSED.
6. For two-stage overhead condensers, do not use reflux from the second stage.
7. Optimize tray deck vapor DP to equal half the weight of liquid on the tray.
8. The weir height should be kept ½" above the downcomer clearance from tray above.
9. Old-style bubble cap trays are more efficient than modern trays, but have 10% to 20% less capacity.
10. Always remember that the reflux comes from the enthalpy, or the heat content, of the reboiler heating medium.

Summary

Saving energy in a distillation tower is not a matter of adding more heat or more reflux. In fact, doing so will often make things worse by causing entrainment and flooding, and reducing fractionation efficiency.

The temperature profile tells you what is really happening inside the tower. If the temperature difference is increasing, separation is improving. If it is decreasing, then you are hurting the tower—no matter what the computer says.

A large portion of the fractionation takes place in the condenser and the reboiler, not just on the trays. Proper control of these, along with the pumparounds, will do far more to improve efficiency than simply increasing tower traffic.

In my experience, the key to saving energy is simple: watch what the tower is telling you, and adjust accordingly.

CONCLUSION

I would like to leave you with the following takeaways.

- Bubble cap trays, which are no longer common, are superior to the perforated (valve) trays used currently, in that bubble caps cannot leak. However, they have 10% to 20% lower capacity than valve trays, and are much more expensive.
- Structured packing has a low pressure drop compared to valve trays and greater capacity. Also, the height needed to develop a theoretical fractionation stage is two-thirds that required for a tray. The problem with structured packing is the potential for poor reflux distribution, which greatly reduces fractionation efficiency.
- Based on experience, I therefore use fixed valve cap trays in new and in retrofit distillation tower designs for fractionation services, with an optimized valve lift for the cap.
- For pumparound (i.e., heat extraction services), I use structured packing, as liquid distribution is less critical. And the capacity and a low pressure drop are of greater importance. Best to use spray nozzles for P/A liquid circulation, with a 100% overlap for the spray pattern.

Troubleshooting

The main tools that I use are:

- Pressure drop surveys,
- Temperature profiles,
- Sample analysis,
- Comparison to computer simulations,
- Radiation scans,

The field pressure drop survey, which I will execute myself, is my most valuable route to troubleshooting distillation problems.

Internal Tower Inspection

- Are all tray deck manways secure?
- Check trays for levelness (not more than 5/8" out-of-level).
- Are weirs level and slightly higher (½") than the bottom edge of the downcomer from the tray above?
- Are downcomer clearances about 2"-2½" and slightly less than the height of the weir?

A Final Word

I once had a very young Process Engineer ask me, in one of my many process troubleshooting seminars, "Mr. Norm, what is the most common distillation problem that you have encountered in your long career?"

Without hesitation, I responded, "It's a process control valve that became stuck out on the unit. Let me explain in detail. It's an excellent example of the critical importance of making personal and direct field observations."

Process Control Valves

"Distillation tower control valves are usually operated by instrument air pressure. A 30 psi air pressure may either close or open a valve, depending on whether the instrument air connection is on top of or underneath the control valve (Figure C.1)," I explained.

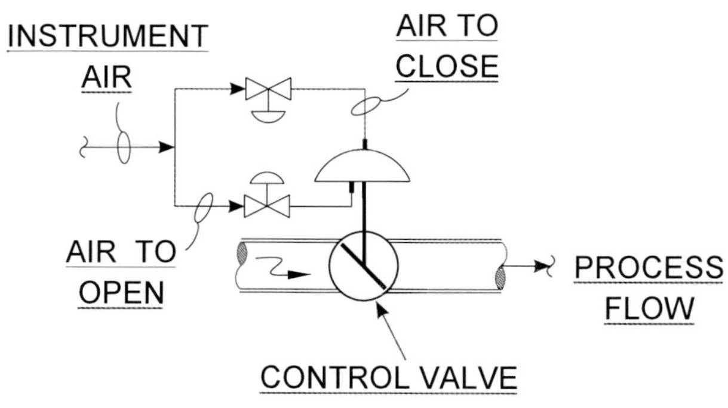

Figure C.1. Control valve action depends on air connection (opens or closes with pressure)

"The most important concept to understand about control valves is that the position of the control valve shown in the control room is NOT the valve position. It is the instrument the air pressure supply to the valve. If the control valve is stuck in a closed position, the valve position in the control room may actually show the valve to be 100% open. It happens all the time!

If you want to determine the real control valve position, look at the control valve position indicator in the field (Figure C.1). Then ask the panel board operator to 'stroke' the control valve, and observe whether the valve actually moves in the field, as it is shown on the control panel."

Direct Field Observations

My comments about control valves also apply to process engineering technology for distillation towers. We have to combine:

1. Our technical knowledge of the theory of distillation.
2. An up-to-date understanding as to how the mechanical components of a distillation tower are designed and assembled.
3. An understanding of the various aspects of the possible malfunctions of both trays and beds of structured packing.
4. Direct field observations as to how the tower internals were installed and combined with analysis of plant operational data.
5. Most important of all, discussion with the experienced plant operators and the maintenance personnel regarding operational problems.

Try to get in touch with reality, which resides in the field, and not in your office.

If you have any questions regarding this text, you can contact me at:

norm@lieberman-eng.com

1-504-430-0757

Thanks for reading my book. I hope it helps you advance in your career.

Norman Preston Lieberman
Chemical Engineer
Cooper Union – class of 1964

GLOSSARY

Bottoms Product. The heavier liquid leaving the bottom of the tower, controlled by reboiler duty and stripping

Chimney Tray. A tray used to collect and redistribute liquid without significant vapor-liquid mixing

Condenser. A heat exchanger that condenses overhead vapor into liquid for reflux and product

Crude Distillation. Separation of crude oil into fractions based on boiling range

Distillation. Separation of liquid mixtures based on boiling point differences

Distillation Tower. A vertical vessel where vapor and liquid interact to separate components

Downcomer. A channel carrying liquid from one tray to the tray below

Dumping. Liquid falling through the tray decks

Entrainment. Liquid droplets carried upward by vapor flow

Energy Efficiency. Achieving separation with minimal heat input

Flooding. Vapor flow prevents liquid from flowing down properly

Fouling. Deposits that reduce efficiency and increase pressure drop

Fractionation. The degree of separation between components

Heat Duty. Amount of heat added or removed in a process

HETP. Height Equivalent to a Theoretical Plate

Liquid Traffic. Downward liquid flow in the tower

Loss of Seal. Vapor displacing liquid, due to low liquid level

Maldistribution. Uneven distribution of liquid or vapor

Non-Condensables. Gases that do not condense in the condenser

Packing. Material providing surface area for vapor-liquid contact

Partial Condensation. Only part of vapor condenses to improve separation

Pressure Drop. Loss of pressure through the tower

Pressure Profile. Pressure variation through the tower

Pumparound. Circulating liquid used for heat removal

Reboiler. Provides heat to generate vapor

Rectification. Enriching vapor in lighter components

Reflux. Tower top product returned to improve fractionation efficiency

Reflux Drum. A vessel that collects condensed overhead vapor and splits it into reflux and product

Relative Volatility. Ratio of the vapor pressures of the light, to the heavy key components

Separation Efficiency. Effectiveness of separation

Stripping. Removal of light components from liquid

Temperature Profile. Temperature variation through the tower

Thermosyphon Reboiler. Uses density differences for circulation

Tower Loading. Operating capacity of tower

Tray. Device for vapor-liquid mixing

Tray Efficiency. Effectiveness of a tray

Turnaround. Scheduled shutdown for maintenance

Vapor Pressure. Pressure exerted by a liquid at a given temperature

Vapor Traffic. Upward vapor flow

Weeping. Liquid leaking through trays due to low vapor flow

Wet Bulb Temperature. Temperature at which a pure liquid evaporates at atmospheric pressure

INDEX

A

B

C

D

E

F

P

R

S

T

V

W

ABOUT THE AUTHOR

Norman Lieberman received the American Institute of Chemical Engineers award for "Practical Art of Application of Distillation Technology in Chemical and Refinery Plants" in 2023. He also received the 2018 "Lifetime Achievement Award" from *Hydrocarbon Processing Magazine.*

Norm graduated from Cooper Union (New York) in 1964, with a degree in Chemical Engineering. He then worked for American Oil until 1980, as a Process Design Engineer, Operating Supervisor of the world's largest alkylation unit, and of refinery process coordination for crude, delayed coking, vacuum distillation, and sulfur recovery.

He then worked for the Good Hope Refinery in Louisiana from 1980 to 1984, initially as Technical Service Manager, Plant Manager, and Vice President, until the refinery was shut down. From 1984 to 1986, he continued to work for Good Hope, troubleshooting 300 partially depleted natural gas wells in south Texas.

In 1987, Lieberman started "Process Improvement Engineering," devoted to:

1. Field troubleshooting refinery process problems.
2. Design and retrofit of:
    a. Crude distillation units,
    b. Vacuum towers,
    c. Delayed cokers,
    d. Sour water strippers,
    e. FCU distillation towers,
    f. Amine absorbers and regenerators,
    g. Alkylation units,
    h. Sulfur recovery.

He has published 11 technical books, and he leads seminars and webinars. A large percentage of worldwide refinery management personnel have attended Norm's seminars since he began sharing his expertise in 1983. About 23,000 engineers and operators have attended 1,020 classes dealing with refinery and process operations.

Norm is married to Elizabeth, a Chemical Engineer, former employee of Mobil Oil (U.K.), and coauthor of *A Working Guide to Process Equipment*, 5th edition, McGraw Hill. They field troubleshoot refinery problems as a team.

Norm won the Rock-n-Roll half-marathon in New Orleans, in the 70 to 79 age group, and came in fifth in his age group in the full marathon. At age 84, he still runs three miles every day.

After 60+ years of process engineering design and field troubleshooting, Norm plans to never stop working and running.